D1622524

POWER
SUPPLIES

**Projects
for the Hobbyist
and Technician**

■■■■■■■■■■■■■■■■■■■■■■■■

David Lines

PROMPT.
PUBLICATIONS

An Imprint of
Howard W. Sams & Company
Indianapolis, Indiana

FIRST EDITION, 1992

PROMPT® Publications is an imprint of Howard W. Sams & Company,
2647 Waterfront Parkway, East Drive, Indianapolis, IN 46214-2012.

This book was originally developed and then published as
Building Power Supplies *by:*
    Master Publishing, Inc.
    14 Canyon Creek Village MS31
    Richardson, Texas 75080
    (214) 907-8938

International Standard Book Number: 0-7906-1024-8

Edited by: *Jerry Luecke, Charles Battle*
Text Design and Artwork by: *Plunk Design, Dallas, TX*
Cover Design by: *Sara Wright*

Acknowledgements
All photographs not credited are either courtesy of Author, Master Publishing, Inc., or Howard W. Sams & Company.

Printed in the United States of America

9 8 7 6 5 4 3 2 1

Table of Contents

Preface

All electronic equipment requires a source of power—a power supply. A power supply can be a separate assembly, or integrated into an electrical or electronic system. It supplies load current at a particular voltage or voltages to the system's circuits. In most cases, a power supply is required to control its output voltage (or voltages) to within close limits as its input voltage and/or output load are changed. This is a regulated power supply. If the output were not regulated, the variations in its output voltage could become signals within the system circuits being powered and cause errors, distortions, extra signals, etc.

Power Supplies has two goals. First, to help you, the reader, understand the basic function of each of the components in a power supply and how the components work together to function as a power supply. Second, to show how to build useful working power supplies. Easy to understand illustrations visually enhance the learning.

Power Supplies begins with a refresher on basic sources of dc and ac power. If you already know these basic principles, you may bypass this material and start with Chapter 2.

Chapter 2 describes the functions of transformation, rectification, and filtering that are required to convert common ac line voltage to a dc voltage. It explains the components needed and identifies the important parameter specifications. The system output is an unregulated voltage.

Chapter 3 begins a pattern that continues through Chapter 6: The basic principles of a system regulating a dc voltage are discussed, and then instructions are presented for building useful power supplies. Chapters 3 and 4 deal with power supply circuits that operate continuously in the linear region, and Chapters 5 and 6 deal with switching power supplies. Specifications, component selection, parts lists, construction plans, design techniques and operating principles are covered.

Performance measurements of each power supply, a calibration procedure for the switching power supplies, and troubleshooting tips are contained in Chapter 7.

Power Supplies should be a good teaching tool because if you complete it, you should understand power supply system principles, and, if you wish, you can build a useful power supply. Those are our objectives. We hope we have succeeded.

D.L.

CHAPTER 1

Basic Sources of DC and AC Power

INTRODUCTION

Power supplies are the energy sources that are required to operate any electrical equipment, including electronic equipment. All types of electronic equipment—from the simplest AM radio to complex computers; from a child's cassette player to the most sophisticated space electronics—need some kind of power supply.

One goal of this book is to show how to build various types of power supplies—from unregulated supplies to regulated switching supplies. With photographs and other illustrations, *Power Supplies* lists the components to use, shows how to interconnect the components to construct a power supply, and shows performance measurements of the completed assembly.

However, *Power Supplies* has another goal—to help you understand the function of each component and the basic principles that are at work as the components perform their functions. If you already know the basic principles, the material in Chapter 1 can serve as a refresher, or you may chose to skip ahead to Chapter 2.

POWER

What is power? Power is the rate at which energy is used. A common unit of electrical power is the watt. A watt is the use of one joule of energy per second. A watt-hour (the unit you see on your monthly electrical bill) is the supply of one watt for one hour, or one joule per second for 3600 seconds. A joule is the amount of energy that one would expend lifting a kilogram (1,000 grams = 2.2 pounds) about 100 centimeters (about 39 inches, or 3.3 feet).

Power supplies for electronic equipment commonly have a wattage rating as a measure of the energy that can be delivered. Technicians and engineers know that the power supply voltage (in volts) times the current (in amperes) delivered to its load (the circuit to which the power is being supplied) is the watts of power supplied. For example, a power supply delivering 5 volts at 10 amperes to its load is providing 50 watts of power.

DC POWER

The most common type of power supply that all of us recognize is a battery. Switch on your flashlight and the batteries inside supply current through the completed circuit to the light bulb. The current through the filament of the light bulb causes it to glow. Power flows from the battery through the light bulb. If the flashlight uses two 1.5-volt batteries (for a total voltage of 3 volts) and the current supplied to the light bulb is 0.1 ampere, the power supplied is 0.3 watt.

The simple circuit for the flashlight is shown in *Figure 1-1a*. This representation of an electrical circuit is called a schematic diagram. Standard symbols are used on schematic diagrams to represent the various components. Notice that the battery symbol is a combination of alternating short and long parallel lines, with the short line indicating the negative (−) terminal.

Figure 1-1b shows a graphical plot of the current supplied by the battery in the circuit of *Figure 1-1a*. The current is constant against time and always in the positive direction. This current is called direct current (abbreviated as dc) because the current always flows in one direction.

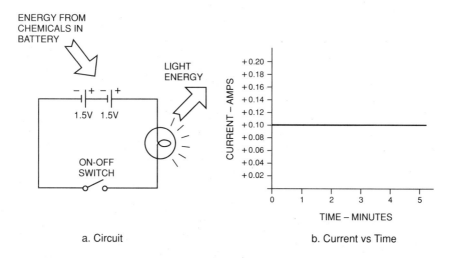

a. Circuit   b. Current vs Time

Figure 1-1. Simple DC Circuit

Battery

A battery is an electrochemical power source. The basic building block of a battery is the electrochemical cell. Higher battery voltages are obtained by connecting several cells together in series. This is how the word "battery" was derived. The cell cross-section shown in *Figure 1-2a* and the schematic shown in *Figure 1-2b* show a battery supplying current to a load of series-connected resistors.

As shown in *Figure 1-2a,* an electrochemical cell contains two plates, called the negative and positive electrodes, and a liquid or semi-liquid between the electrodes, called the electrolyte. When a circuit (conductive path) is completed between the terminals of the electrodes, a chemical reaction takes place between the positive electrode and the electrolyte that "pulls" electrons from the circuit to the positive terminal and into the electrolyte. An opposite reaction takes place at the negative electrode, "pushing" electrons out the negative terminal and into the circuit. Electrons and positive ions flow in the electrolyte to maintain the chemical reaction.

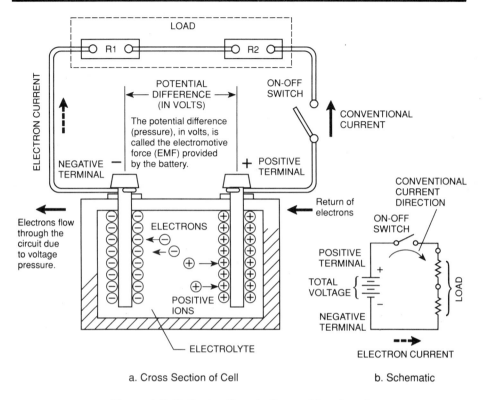

a. Cross Section of Cell

b. Schematic

Figure 1-2. Batteries Supply Current to a Load

In *Figure 1-2a,* two directions are shown for the current in the circuit across the battery. One is electron current; the other is conventional current. Electrons have a negative charge and flow from a point of negative voltage to a point more positive in voltage. Because early scientists who discovered electricity assumed that conventional current was the movement of positive charges (opposite from electrons), the conventional circuit current direction is opposite to the electron current direction.

Conventional current direction is the standard most commonly used by the electrical and electronic industries; therefore, it will be used in the remainder of this book. Just remember, electron current will be in the opposite direction to the conventional current direction shown.

The chemical reaction develops a potential difference, called voltage, between the positive and negative terminals. Voltage is measured in units of volts. The pressure caused by the voltage is called the electromotive force. The number of electrons that actually flow through a conductor per second is called current, and is measured in units called amperes, commonly called amps. The voltage of a battery decreases as the battery supplies current to its load; therefore, batteries are commonly rated in "ampere-hours." This rating specifies the number of hours the battery can supply one ampere of current before the battery voltage decreases to a specified minimum value. For batteries that are designed to supply less than one ampere, this rating is expressed in "milliampere-hours," where one milliampere equals 0.001 ampere.

Primary and Secondary Batteries

Batteries are available in many types and voltages for many different purposes. Some are rechargeable (secondary cells) and some are non-rechargeable (primary cells). An example of a primary battery is the alkaline cell used in everything from flashlights to cameras to portable stereos. Examples of secondary batteries are the nickel-cadmium (Ni-Cd) batteries in some cordless tools, and the lead-acid batteries used in automobiles. Batteries are constantly being improved to increase the power delivered or to reduce the physical size required for a given power. Such developments increase their use in portable electronic equipment.

DC Generator

Another power supply for direct current is a dc generator. To understand the operation of a generator, you must understand the basic principles at work when a wire is moved in a magnetic field.

Generator Principles

Look at *Figure 1-3*. Here a loop of wire is mounted on a shaft and positioned in a magnetic field. The magnetic field, represented by lines in *Figure 1-3*, runs from the North pole of the magnet to the South pole. This field is invisible, but it is present. You know this is so because if a nail is brought close to the magnet it will be pulled to the magnet. The lines are made visible in *Figure 1-3* to illustrate the principle. Notice that a voltmeter is connected across the ends of the wire loop. The loop's 90 degree position is as shown—the loop sides are closest to the pole pieces, and the face of the loop is parallel with the magnetic field.

Now, if the loop starts at 0 degrees and is rotated rapidly clockwise 180 degrees, the voltmeter needle will deflect to indicate a momentary voltage and then fall back to zero. When the loop *cut* across the magnetic field lines, an electromotive force was generated in each side of the wire loop to cause current in the same direction through the meter circuit and deflect the meter. The amplitude of the generated voltage depends on three factors: the strength of the magnetic field, the speed of the rotating loop, and the number of turns of wire on the loop. Increasing any factor increases the generated voltage.

If the loop is again rotated clockwise another 180 degrees, the voltmeter needle will again deflect, but in the opposite direction. However, if before the second 180-degree turn, the meter leads were reversed, the meter needle would deflect in the same direction as for the first rotation. Thus, if the loop is continually rotated while the leads are reversed at each 180-degree point, the generated voltage plotted against time will look like the waveform shown in *Figure 1-4a*. This is the basis of a direct-current generator.

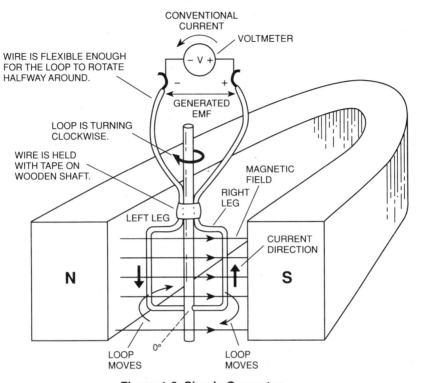

Figure 1-3. Simple Generator

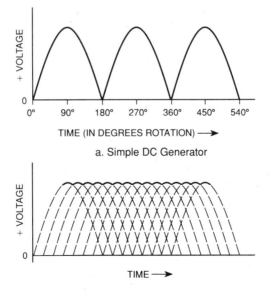

a. Simple DC Generator

b. Commercial DC Generator

Figure 1-4. DC Generator Voltage Versus Time

Commercial Generator

A commercial dc generator has many windings, each with many turns of wire, wound on an iron core. This assembly is called an armature. The ends of the windings connect to split slip rings, called the commutator, on the generator shaft. The output terminals are connected to sliding connectors, called brushes, which are in contact with the commutator. This combination reverses the winding connections to the output terminals as the armature rotates so that the generated voltage is always the same polarity. The windings are positioned on the generator shaft so that the output voltage waveform looks like that shown in *Figure 1-4b* when plotted against time. The output voltage is nearly constant and when connected to a circuit load, the current in the circuit will flow in only one direction. It will be direct current from a dc generator—a dc power supply.

As mentioned above, the voltage level produced by a generator depends on the number of turns in the generator windings, on the speed at which the generator shaft is driven, and on the strength of the magnetic field. The maximum current that a dc generator can produce depends on the size of the wire in the generator windings, the design of the commutator and brushes, and the way the generator is cooled. The rated power wattage is equal to the rated voltage times the rated current.

Induction

The voltage in the loop that caused a current in the meter connected to the loop is said to be induced by the magnetic field. The induction can occur either by moving the loop through the magnetic field, or by moving the magnetic field through the loop. The principle of induction is very important to the dc generator, to the ac generator discussed next, to inductors, and to transformers discussed in Chapter 2.

AC POWER

Alternating current (ac) differs from dc in that the circuit current doesn't flow only in one direction; *it reverses and flows in the opposite direction as well.* It changes direction at regular time intervals. The regular periodic rate at which the current reverses direction is called the *frequency.* AC-powered equipment must be able to operate at the frequency at which the alternating current is generated.

AC power runs trains, factories, and the appliances in our homes. What makes it especially useful is that transformers can be used to convert ac to different voltages, as we will see in Chapter 2.

AC Generator

The primary source of ac power is the ac generator, which is usually called an alternator. It is very similar to the dc generator. Look again at *Figure 1-3.* Recall that when the loop coil was rotated the first 180 degrees, the meter deflected in one direction; and if continued through a second 180 degrees without reversing the leads, the meter deflected in the opposite direction. The change in direction of the meter's deflection meant that the current changed its direction through the meter. The voltage induced in the loop was in the opposite direction so the current in the circuit changed direction.

Look at *Figure 1-5a*. The shaft on which the loop of wire of *Figure 1-3* is mounted is now driven by a motor that turns the loop so it makes 60 complete rotations each second. The ends of the loop are connected to slip rings so the meter is connected to the loop through brushes on the slip rings at all times as the loop turns. The voltage indicated by the meter as the loop turns is shown in *Figure 1-5b*. Notice that in the first 180 degrees, the voltage is positive, and in the next 180 degrees, the voltage is negative.

As the loop turns, the voltage repeats itself each 360 degrees, or each cycle. If the loop makes one 360 degree rotation in 1/60 of a second, then 60 cycles of voltage will be induced or generated in one second, and the frequency of the ac voltage is 60 cycles per second. The official name for "cycles per second" is "hertz," so the generated voltage has a frequency of 60 hertz. If the number of turns on the loop of wire is increased until the generated voltage is 110 volts, then a 110VAC, 60Hz ac generator will have been constructed. This is the principle of an ac generator—the primary ac power supply. Because of its simpler construction (continuous slip rings instead of a split commutator), and because dc is easily derived from ac, as we will see in Chapter 2, the ac generator or alternator is much more widely used than the dc generator.

Inductance

Let's review up to this point. A wire *cutting across* a magnetic field will have a voltage induced in it and there will be a current in the wire when it is connected in a complete circuit. There is a counter principle to this as shown in *Figure 1-6a*. When there is current in a wire, a magnetic field will be formed around the wire as shown. As current increases in the wire, the magnetic field increases around the wire; as current decreases in the wire, the magnetic field decreases around the wire. If other wires are near the wire carrying the varying current, as shown in *Figure 1-6b,* the varying magnetic field *cuts across* the wires and induces a voltage in them. The varying magnetic field expanding and collapsing around the nearby wires is the same as if the magnetic field were stationary and the nearby wires where *cutting across* the magnetic field. This is the principle of the transformer which will be discussed further in Chapter 2. Inductors also use this principle to produce an electrical property called inductance.

Inductive Reactance

When current passes through wire which is wound in a coil, a magnetic field will be produced around the coil which is the result of the interactions of the magnetic fields around each loop of wire in the coil. With alternating current through the coil, the field expands and collapses around the coil. The expanding and collapsing field induces a counter voltage in the coil that opposes the

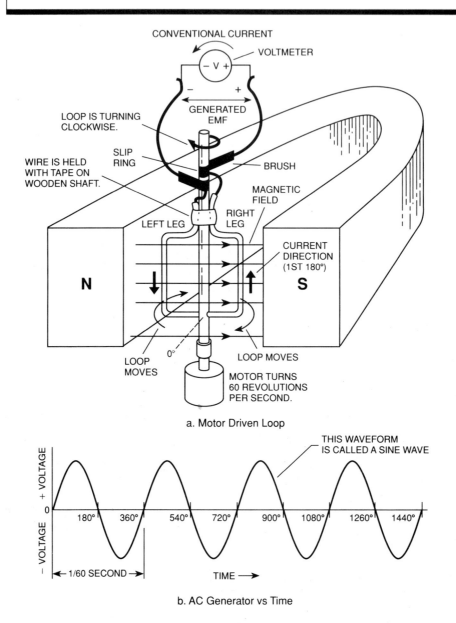

a. Motor Driven Loop

b. AC Generator vs Time

Figure 1-5. AC Generator

varying current that is producing the original field. This opposition or resistance to ac is called inductive reactance. It has the symbol X_L and is calculated using the following equation with f as the frequency in hertz and L as the inductance of the coil in henrys; π is a constant equal to 3.1416:

$$X_L = 2\pi f L$$

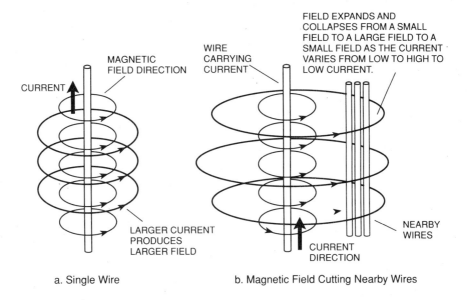

Figure 1-6. Magnetic Field Around Wires Carrying Curent

Inductance depends on the size (area and length) of the coil, the number of turns, and the permeability of the material on which the coil is wound. Permeability is a measure of how easily a material can be magnetized. Notice that as the frequency of the varying current increases, the inductive reactance increases. This will be important later when inductors used in power supply filters are discussed.

The unit of inductive reactance is ohms, which is the same unit used for the resistance of a resistor. The total opposition (impedance) by an inductance in an ac circuit is the sum of the inductive reactance and the dc resistance of the wire in the inductor. An inductor with dc or very low frequency ac flowing through it has an impedance close to the dc resistance of the wire used to wind it. As the frequency increases, the impedance increases dramatically because the inductive reactance increases dramatically.

SUMMARY

In this chapter, we have discussed the basic sources of dc and ac electricity — electrochemical (batteries) and electromechanical (generators, alternators). We have illustrated how magnetic principles are used in generators and alternators to generate electricity and in inductors to produce inductive reactance, an electrical property that opposes alternating current. In Chapter 2, we will see how the magnetic principles are used in transformers, one of the main components in unregulated power supplies.

CHAPTER 2

Unregulated Power Supply Systems

INTRODUCTION

Most electronic equipment circuits require a power supply that provides a dc voltage at some maximum dc current. Batteries can be used to provide this energy for equipment that has a low power requirement or if it is normally used intermittently. Another possible power source is a dc generator, however, neither batteries nor dc generators are practical or economical for many types of electronic equipment. Since the 110VAC, 60-hertz power distributed by the electric power company is readily available, most electronic equipment operates from a power supply that converts the ac line voltage to a dc voltage.

The basic unregulated dc power supply functions are *transformation*, *rectification*, and *filtering*. These functions are indicated in *Figure 2-1*, along with the input and output waveforms for each function.

The *transformation* function input is the 110VAC from the utility line. Its ac output voltage, which can be lower or higher than the input voltage, is the input to the rectification function. The *rectification* function output is a dc voltage, but because it has large amplitude variations, it is called "pulsating dc." The *filtering* function reduces the large amplitude variations so that the

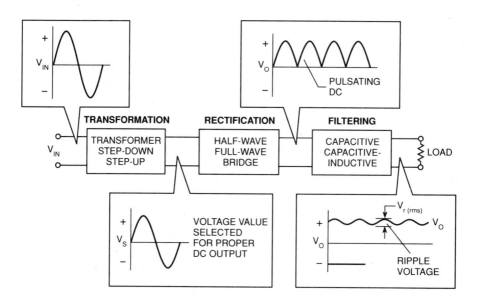

Figure 2-1. Unregulated DC Power Supply

output is a dc voltage with only a small "ripple" voltage riding on it. This basic power supply is called an unregulated dc power supply because its output varies with changes in the ac input voltage as well as changes in the load on the power supply output. To understand the operation of this power supply, let's discuss each of the functions in detail.

TRANSFORMATION

Transformation provides two primary functions:
- It changes the line voltage value to the voltage value required to produce the proper dc voltage output.
- It electrically isolates the electronic equipment from the utility power line.

Transformers

A transformer is the component that performs the transformation function. As shown in *Figure 2-2*, it consists of at least two coils of wire wound on the same iron core. The coil of wire receiving the input voltage is called the primary; the coil providing the output voltage is called the secondary. Many times there are two or more secondaries. The basic operating principle of a transformer is induction, which was discussed in Chapter 1. Varying current from an ac voltage applied to the primary creates a changing magnetic field in the iron core. This magnetic field, coupled to the secondary through the core, cuts the secondary windings and induces an ac voltage in each turn of the secondary. Thus, energy is transferred from the primary to the secondary by the varying magnetic field without any electrical connection between them.

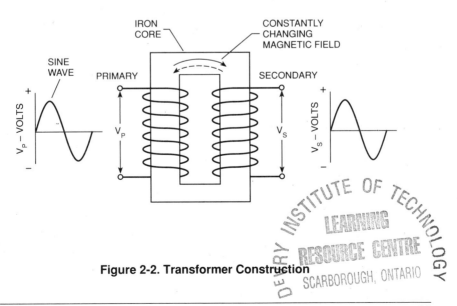

Figure 2-2. Transformer Construction

Isolation

Because the energy transfer is accomplished only by the magnetic coupling between the primary and the secondary, the secondary and any circuits connected to it are isolated from the primary and any circuits connected to it. This is important for safety because the primary is connected to the high current supply of the utility line. Without such isolation, there is a serious shock hazard. Another advantage is that no dc connection exists between circuit ground in the primary circuit and circuit ground in the secondary circuit.

Turns Ratio

In an unregulated power supply, the transformation function must provide the ac output voltage value required to produce the proper dc output voltage. This is easily accomplished in a transformer by varying the ratio of the number of secondary turns (N_s) to the number of primary turns (N_p). The ratio is formed by dividing N_s by N_p; that is, N_s/N_p. The secondary voltage can be less than or greater than the primary voltage just by varying the turns ratio, N_s/N_p.

The amount of voltage induced in a turn of the secondary winding is the same as the voltage induced in a turn of the primary. The voltage induced in each turn of the primary, e_p, is the primary voltage, V_p, divided by the primary turns, N_p. In equation form, it is:

$$e_p = V_p/N_p$$

If the same voltage is induced in each turn of the secondary winding, then the secondary voltage, V_s, is the number of secondary turns, N_s, times the induced voltage, e_p. In equation form, it is:

$$V_s = N_s \times e_p$$

By substituting the value of e_p from the first equation into the second equation, V_s equals the turns ratio times V_p. In equation form, it is:

$$V_s = N_s/N_p \times V_p$$

Knowing that the secondary voltage is the primary voltage times the turns ratio, it is easy to see how the transformer can be used to vary the voltage level of the ac voltage used in the unregulated power supply system. Let's look at a couple of examples using a schematic symbol for a transformer like that shown in *Figure 2-3a*.

Step-Down Transformer

If the secondary turns are fewer than the primary turns, the secondary voltage is less than the primary voltage. This is a step-down transformer. In the example shown in *Figure 2-3b*, the turns ratio is 0.5 and the primary voltage is 110VAC; therefore, the secondary voltage is 110VAC × 0.5 = 55VAC.

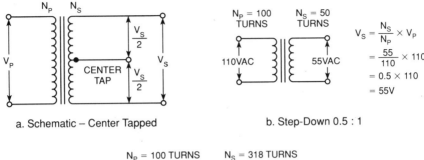

a. Schematic – Center Tapped

b. Step-Down 0.5 : 1

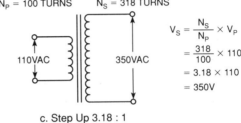

c. Step Up 3.18 : 1

Figure 2-3. Transformers

Step-Up Transformer

If the secondary turns are more than the primary turns, the secondary voltage is greater than the primary voltage. This is a step-up transformer. In the example shown in *Figure 2-3c*, the turns ratio is 3.18 and the primary voltage is 110VAC; therefore, the secondary voltage is 110VAC × 3.18 = 350VAC.

Power Transfer and Efficiency

The power flowing out of the secondary in relation to the power flowing into the primary is given by:

$$V_p \times I_p = V_s \times I_s \times n$$

where n is the transformer efficiency. If the efficiency were 100% or n = 1, the power out would be equal to the power in. Therefore, if V_s is smaller than V_p, I_s must be larger than I_p, and if V_s is larger than V_p, I_s must be smaller than I_p. A power transformer's efficiency usually is between 85% and 95%.

AC VOLTAGE VALUES

A dc voltage usually has only one measured value. An ac voltage, because it is continually changing, can have several different measured values depending on how it is measured. Study these voltage values because they are important in power supply design. They will be used later in this book for the project designs.

Figure 2-4 shows the common ac power-line voltage waveform plotted against time. The time axis is also calibrated in the degrees of rotation of the ac cycle. For the first 180 degrees, the voltage is positive; for the next 180 degrees, it is negative. Because the power-line frequency is 60 hertz, a 360 degree rotation cycle takes one sixtieth of a second.

V_{rms}

The value of the power-line voltage is typically stated as 110VAC, but this value is really $110V_{rms}$. V_{rms}, as shown in *Figure 2-4*, is the value normally measured on an ac voltmeter. V_{rms} is the usually stated value of an ac voltage because it is an ac voltage that provides the same energy to a resistive load as an equivalent dc voltage.

V_{pk} and V_{pp}

Note in *Figure 2-4* that other values can describe the ac voltage. The peak voltage, V_{pk}, is the maximum value of the voltage in its cycle. The peak-to-peak voltage, V_{pp}, is the total value of the voltage from the maximum positive peak to the maximum negative peak. For a sine-wave voltage (which is a voltage that varies as the mathematical sine function of the angle of rotation) as shown in *Figure 2-4*, V_{rms} equals 70.7% of the peak voltage ($V_{rms} = 0.707 \times V_{pk}$). Conversely, V_{pk} is 141.4% of the V_{rms} voltage ($V_{pk} = 1.414 \times V_{rms}$). The peak-to-peak voltage for a sine-wave is twice the peak voltage ($V_{pp} = 2 \times V_{pk}$).

RECTIFICATION

Rectification converts an ac voltage into a dc voltage. The component that performs the rectification function is called a rectifier.

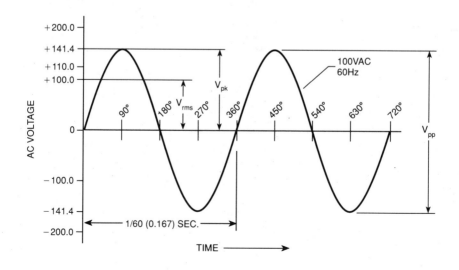

Figure 2-4. AC Voltage Values

Rectifier

The rectifier is a type of diode. Simply stated, it is a one-way valve for electricity. As indicated in *Figure 2-5*, it allows electrons to flow freely in only one direction, called the forward-biased direction, where the anode voltage is more positive than the cathode voltage. In the opposite direction, called the reverse-biased direction, electrons cannot flow easily. In this direction, the anode voltage is more negative than the cathode voltage.

V_F

A forward-biased diode is not like a piece of wire because it has enough resistance to produce a significant voltage drop across it. The forward-biased voltage drop, shown as V_F in *Figure 2-5a*, varies according to the type of material used in the diode. A commonly used silicon diode has a V_F of 0.5-0.7V, and a germanium diode has a V_F of 0.2-0.3V. V_F varies from 0.05V to 0.2V with large changes in I_F.

V_R and PIV

The maximum reverse-bias voltage, shown as V_R in *Figure 2-5b*, that can be applied to a diode is called the peak inverse voltage (PIV). If this voltage is exceeded, the anode-cathode junction may break down and allow a large current to flow in the reverse direction. If breakdown is exceeded, the diode is usually permanently damaged.

Half-Wave Rectifier

A simple half-wave rectifier circuit has a diode connected in series with a transformer secondary output as shown in *Figure 2-6a*. The input primary voltage is a power-line, 60-hertz, sine-wave voltage. The positive cycle alternation is labeled A and the negative one is labeled B. The polarities of the primary and secondary voltages are noted for each alternation.

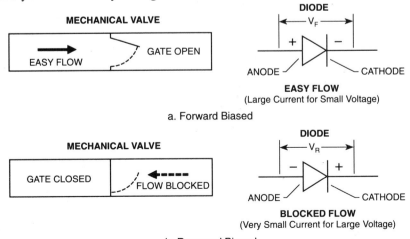

Figure 2-5. Diode: A One Way Valve for Current

On alternation A, diode D1 conducts because its anode is more positive than its cathode. A voltage equal to the secondary voltage (minus the diode's V_F) is developed across the load, R_L. On alternation B, D1 blocks current so no voltage is developed across R_L. The secondary voltage appears as reverse voltage, V_R, across the diode. To withstand this voltage, D1's PIV must be greater than the secondary voltage's V_{pk}.

The output voltage *(Figure 2-6a)* is a series of 60-hertz, half-cycle alternations of the secondary voltage. The voltage is always in one direction and is known as "pulsating dc." If the area under the positive pulses is averaged over a complete cycle, the average dc voltage is 0.318 times the secondary's V_{pk}.

A half-wave rectifier circuit is used for low current applications such as battery chargers and ac-dc adapters for calculators. Note that if a battery were placed across the dc output as a load, it would charge to the peak voltage of the secondary minus the V_F of the diode.

Full-Wave Rectifier

The rectifier circuit in *Figure 2-6b* converts both alternations of the secondary voltage to a dc voltage; therefore, it is known as a full-wave rectifier. A diode is in series with each secondary output and the center tap is grounded. The voltage between the center tap (See *Fig.2-3a*) and each secondary output is equal in value, but opposite in direction. When V_{S1} is positive, V_{S2} is negative.

D1 conducts during the A alternation and D2 conducts during the B alternation. The center tap is a common return for the current from each diode. Because both diodes conduct current to the load in the same direction, the pulsating dc has positive half-cycles on both alternations of the secondary voltage. The output is pulsating dc at 120 hertz with an average dc voltage of $0.636V_{pk}$.

Only half of the secondary is used at one time; therefore, the transformer secondary output voltage must be twice that needed to provide the proper dc voltage. Also, each diode's PIV must be at least the full secondary's V_{Spp}.

Bridge Rectifier

The rectifier circuit of *Figure 2-6c* is called a full-wave bridge rectifier. It uses four diodes in a bridge network. One output terminal of the bridge network is a common ground for the return of the load current. The other output terminal is connected to the load.

D1 and D2 conduct on alternation A, and D3 and D4 conduct on alternation B. Both conducting paths deliver current to the load in the same direction. The pulsating dc output is the same as the full-wave rectifier. The output dc voltage is the secondary voltage minus two forward diode drops. The diodes' PIV must be greater than the secondary peak voltage, V_{Spk}.

FILTERING

The pulsating dc output after transformation and rectification is not a satisfactory power source for most electronic circuits. The filtering function smooths the output so that a nearly constant dc is available for the load.

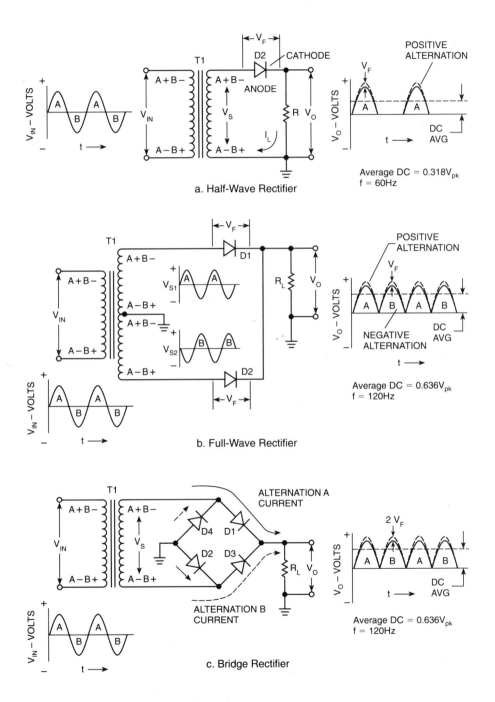

a. Half-Wave Rectifier

b. Full-Wave Rectifier

c. Bridge Rectifier

Figure 2-6. Rectifier Circuits

The pulsating dc output from the rectifier contains an average dc value and an ac portion called a ripple voltage. A filter circuit reduces the ripple voltage to an acceptable value. Resistors, inductors and capacitors are used to build filters. None of these components have amplification. Resistors oppose current and normally function the same way in dc or ac circuits by not varying with frequency. Inductors oppose current changes, and their inductive reactance increases with frequency. Capacitors oppose voltage changes, and their capacitive reactance decreases with frequency. Let's take a closer look at capacitors and understand what they do.

Capacitor

A capacitor consists of two conductive plates separated by an insulator, called the dielectric, as shown in *Figure 2-7a*. When a dc voltage is applied across the plates, electrons collect on one plate and positive ions on the other. The difference in the electrical charges on the plates equals the voltage applied. If the voltage is removed, the electrical charges remain in place and maintain the voltage difference between the plates. In other words, the charge is stored by the capacitor. The charge storage characteristic gives a capacitor in a circuit the effect of opposing voltage changes. This effect is very important to the filtering function in dc power supplies.

The electrical unit of capacitance is the farad. A farad is a very large quantity so actual capacitors are usually rated in microfarads. One microfarad is 0.000001 (1×10^{-6}) farad.

Discharging a Capacitor

To understand how a capacitor is used in filtering, let's examine a capacitor's discharge characteristic. In *Figure 2-7c*, a capacitor has already been charged to a voltage V_c. Switch S, which has been open, is closed and the capacitor discharges through resistor R. The voltage across the capacitor decreases as time passes according to the very predictable curve shown in *Figure 2-7d*. It is called an RC discharge curve because the time scale is in RC time constant units. To find the RC time constant (in seconds) for the discharge curve, multiply the resistance (in ohms) discharging the capacitor by the capacitance (in farads). If the capacitor is 10 microfarads (0.00001 farad) and the resistance is 100 ohms, one RC time constant is 0.001 second ($100 \times 0.00001 = 0.001$). The resistance discharging a power supply filter capacitor is the power supply load.

The discharge curve of *Figure 2-7d* shows that V_c will decrease to 37% of its original charged value in one RC time constant. In five RC time constants, the capacitor will be fully discharged. Analysis of the discharge curve reveals two important facts:

- The larger the capacitance, the larger the RC time constant and the slower the discharge.
- The smaller the resistance, the smaller the RC time constant and the faster the discharge.

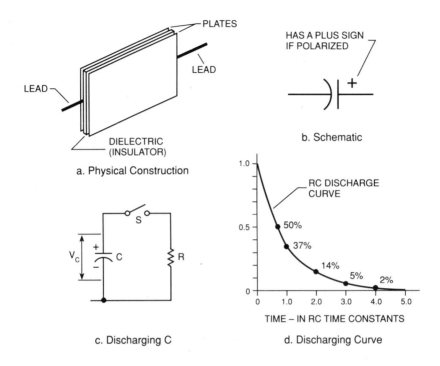

Figure 2-7. Capacitors

Capacitive Filter

The simplest filter is a single capacitor in parallel with the output from a rectifier. In *Figure 2-8a*, C_{F1} is the filter capacitor and R_L represents the power supply load. I_L equals V_o divided by R_L (Ohm's law). Look at the waveform of V_o plotted against time. V_o increases rapidly to the peak voltage output from the rectifier as the first alternation half-cycle charges C_{F1}. If there were no load (R_L=infinity), V_o would remain at the peak voltage and C_{F1} would not have to be charged again; however, with a load of R_L and a current of I_L, C_{F1} starts to discharge as the rectified pulse decreases to zero. The discharge is according to the curve of *Figure 2-7d*, and the RC time constant is C_{F1} times R_L. C_{F1} discharges to point A shown on the V_o waveform. At point A, the next alternation pulse voltage increases above V_o and recharges C_{F1} to the peak voltage again.

A filter capacitor must be large enough to store a sufficient amount of energy to provide a steady supply of current for the load. If the capacitor is not large enough, or is not being charged fast enough, the voltage will drop as the load demands more current.

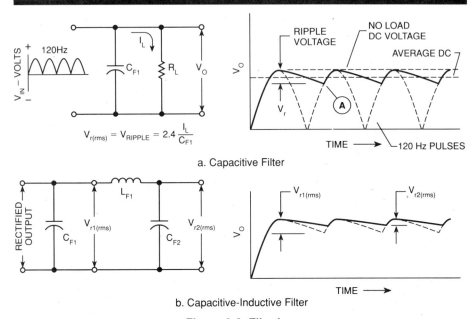

a. Capacitive Filter

b. Capacitive-Inductive Filter

Figure 2-8. Filtering

The RMS voltage of the variation in V_o above and below the output dc value is shown as the ripple voltage, v_r, in *Figure 2-8a*. For a full-wave, 60Hz ac input, $v_{r(rms)}$ is:

$$v_{r(rms)} = 2.4\ I_L/C_{F1}\ ^{(1)}$$

where I_L is in milliamperes, C_{F1} is in microfarads, and an RC discharge is assumed.

The equation can be rearranged to find C_{F1} for a power supply that must supply I_L milliamperes of current and have a ripple voltage of $v_{r(rms)}$.

$$C_{F1} = 2.4\ I_L/v_{r(rms)}\ ^{(1)}$$

Ripple Voltage

Ripple voltage is usually stated as a percent of the dc output voltage, V_o:

$$\%v_{r(rms)} = v_{r(rms)}/V_o \times 100$$

For example, let's assume a power supply must provide 10V dc at 200 milliamperes with a 1% maximum ripple voltage. Using the above equation, multiply 1 times 10 and divide by 100 to find that the ripple voltage must not exceed 0.1 volt. Then use the C_{F1} equation to find that C_{F1} must be 4800 microfarads to meet the ripple requirement.

Additional filter components can be added, as shown in *Figure 2-8b*, to reduce the ripple voltage further. When L_{F1} is in henries, C_{F2} is in microfarads, and the source is a 60Hz, full-wave rectifier, the $V_{r2(rms)}$ of *Figure 2-8b* is:

$$v_{r2(rms)} = v_{r1(rms)} \times 1.77/L_{F1}C_{F2}\ ^{(1)}$$

[1]*Basic Electronic Technology*, Radio Shack No. 62-1394 © 1985 Texas Instruments, Dallas, TX.

VOLTAGE DOUBLER

A transformer can increase the voltage level of ac, but a transformer is relatively expensive and increases the weight and heat generation of a power supply. In some cases, a special type of rectifier circuit called the voltage doubler is used to provide a higher output voltage without using a larger and heavier step-up transformer.

Operation

Figure 2-9a illustrates the circuit and *Figure 2-9b* shows the output waveform. The first A alternation forward biases D1 and charges C1 to V_{pk}. The first B alternation forward biases D2 and charges C2 to V_{pk}. Because C1 and C2 are connected in series with their polarities aiding, the output voltage, V_o, is the addition of the two capacitors voltages, which is $2 \times V_{pk}$. Additional circuit segments can be used to form voltage triplers and quadruplers. Circuits such as this are useful for photo-flash power supply applications that require a high-voltage, high-current pulse for a short time period, and then allow a relatively long time before another pulse is required.

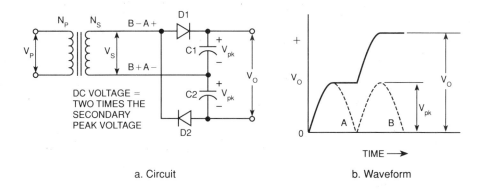

a. Circuit   b. Waveform

Figure 2-9. Voltage Doubler

SUMMARY

An unregulated power supply provides the rated output voltage when the rated line voltage is input and the rated current is delivered to the load. If the line voltage increases or if the load current decreases, the output voltage increases. If the line voltage decreases or if the load current increases, the output voltage decreases. Thus, the output voltage is unregulated because it varies with line and load changes.

The basic functions required for unregulated power supply systems of transformation, rectification and filtering were discussed in this chapter. The components that perform these functions were explained and their important parameters were identified. In the next chapter, the circuits required for regulation will be added.

CHAPTER 3

Basic Regulated Power Supply Systems

INTRODUCTION

Modern digital ICs provide many of the conveniences enjoyed today. Most of these ICs require a precise power supply that controls the voltage level within narrow limits. The supply must respond very quickly to peaks and dips in current demand, because the ICs are adversely affected if certain voltage variations occur. This chapter explains the basic methods for controlling the output of the power supply so that changes in load current and input line voltage have little or no effect on the output voltage. A power supply with its output controlled in this fashion is called a regulated power supply.

To design a regulated power supply, an unregulated supply like those described in the last chapter is used with a regulator circuit added to its output. This chapter will show how a regulator can monitor the power supply output and automatically make adjustments so the output voltage stays within defined limits. The basic functional parts of a regulator are discussed along with a variety of integrated circuits that perform these functions. These ICs simplify the design and manufacture of a regulated power supply.

WHY IS A VOLTAGE REGULATOR NEEDED?

Figure 3-1a is a schematic circuit diagram of the unregulated power supply shown in the functional diagram of *Figure 2-1*. *Figure 3-1b* shows a simplified schematic equivalent to the circuit shown in *Figure 3-1a*. It has a certain open-circuit (no load) dc voltage as its source, V_{DC}, and a series resistor, R_Z, called the output impedance. Any load current, I_L, flows through R_Z. Two main factors work to change the output voltage, V_O. One is a change in load current; the other is a change in input line voltage, V_{IN}, which changes V_{DC}. Sometimes these two factors occur independently, and sometimes they interact with each other.

Variations in Load Current

The dc output voltage, V_O, varies as the current demands of the load fluctuate. *Figure 3-2* plots the variations. If there is no load current ($I_L=0$), then V_O equals V_{DC}, the peak ac level from the power transformer. As the load demands more current, the voltage drop across R_Z causes the output voltage to fall. When the load is at the rated output current, I_{LR}, V_O is at V_{OR}, a point near the transformer's rated RMS voltage. This voltage variation is not a problem if the load circuit demands somewhat constant current, or if the load circuit operation can tolerate the variations in V_O. For example, the power supply voltage for an audio power amplifier can vary over a considerable range without affecting the amplifier operation, as long as it doesn't dip below the level where signal clipping or other distortion occurs.

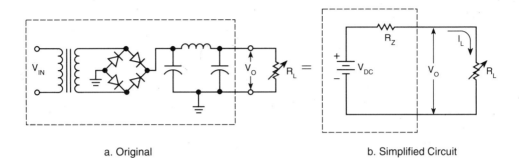

a. Original   b. Simplified Circuit

Figure 3-1. Unregulated Power Supply Circuit

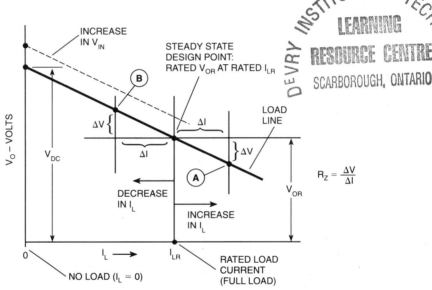

Figure 3-2. Load Line of Unregulated Power Supply

Variations in Line Input Voltage

As shown in *Figure 3-2*, V_O also varies because the line input voltage, V_{IN}, varies. If V_{IN} changes, as it often does in many areas, the power transformer output and the filtered V_O output will change.

VOLTAGE REGULATOR PRINCIPLES

To maintain a constant V_O, a regulator circuit is inserted between R_Z and R_L as shown in *Figure 3-3*. There is a voltage drop across the regulator, V_{REG}; therefore, the input voltage, V_{DC}, must be larger than V_{DC} shown in *Figure 3-1b*. In *Figure 3-3*, $V_O = V_{DC} - (R_Z + V_{REG})$.

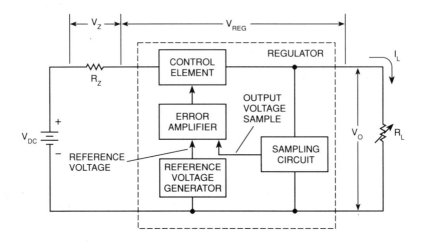

Figure 3-3. Simplified Regulator Circuit

Regulator Action

To perform the required regulation, the regulator circuit varies V_{REG} to keep V_O constant as I_L and V_{DC} change. If I_L increases, V_Z increases which tends to reduce V_O; however, the regulator reduces V_{REG} to offset the increase in V_Z so V_O remains constant. Conversely, if I_L decreases, which tends to increase V_O, the regulator increases V_{REG} to keep V_O constant. Similarly, the regulator increases or decreases V_{REG} if V_{DC} increases or decreases, respectively.

Sampling Circuit

The sampling circuit monitors the output voltage and feeds an output voltage sample to the error amplifier.

Reference Voltage Generator

The reference voltage generator maintains a constant reference voltage for the error amplifier regardless of input voltage variations.

Error Amplifier

The error amplifier compares the output voltage sample to the reference voltage and generates an error voltage if there is any difference between them. The error amplifier output is fed to the control element to control the value of V_{REG}.

Control Element

The control element is essentially a variable resistance which is in series with V_{DC}, R_Z and R_L. When V_{DC} or I_L changes, the input from the error amplifier adjusts this variable resistance to change V_{REG} to hold V_O constant as explained above.

Now that the "big picture is understood," let's learn the details of the regulator circuit operation. First, the basic transistor action which is used in the circuits will be discussed.

TRANSISTOR ACTION

Transistor Construction

To clarify what is meant by transistor action, look at *Figure 3-4*. It is divided into three parts. *Figure 3-4a* shows the construction of an integrated circuit NPN transistor. The transistor is made up of islands of N, P and N semiconductor material diffused into a semiconductor wafer substrate. The most commonly used material is silicon. On the substrate, the devices next to each other are isolated from each other so they can act independently. Small gold wires are bonded to defined evaporated metal contacts to the N, P and N regions to make external electrical connections to the base (B), emitter (E) and collector (C)—the three terminals of a transistor.

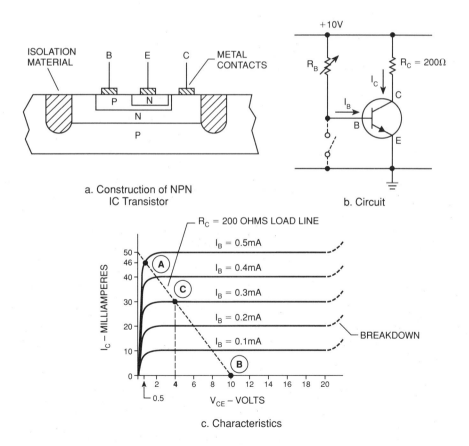

a. Construction of NPN
IC Transistor

b. Circuit

c. Characteristics

Figure 3-4. Transistor Construction and Operation

Transistor Circuit and Characteristic Curves

Figure 3-4b shows a transistor in a circuit and *Figure 3-4c* describes transistor action with a plot of collector current, I_C, against the collector-to-emitter voltage, V_{CE}. Base current, I_B, flows from base to emitter when the base is more positive than the emitter by a base-to-emitter voltage, V_{BE}, greater than 0.7V. With a given base current, a characteristic curve of collector current can be plotted as V_{CE} is varied. Different characteristic curves are shown in *Figure 3-4c* for different base currents. For example, if I_B equals 0.1 milliampere, then I_C is approximately 10 milliamperes for any V_{CE} from 2 volts to 20 volts.

Notice that there is a current gain in the transistor. I_C is 10 milliamperes for an I_B of 0.1 milliamperes; therefore, the base current controls a collector current that is 100 times greater. The current gain is not affected much by V_{CE} until V_{CE} is less than 2 volts or greater than the collector-to-emitter breakdown voltage.

Load Line

If R_C in *Figure 3-4b* is 200 ohms and if the supply voltage is 10V, then for any base current, I_B, the V_{CE} of the transistor will fall on the diagonal dashed line shown on the characteristic curves of *Figure 3-4c*. This line is called a load line. If $I_B=0$, then $I_C=0$ and the operating point would be at point B on the load line where V_{CE} is 10 volts. If $I_B=0.5$ milliampere, then $I_C=46$ milliamperes, which is at point A on the load line. If $I_B=0.3$ milliampere, then $V_{CE}=4$ volts, which is at point C on the load line. Thus, as I_B varies, I_C changes and V_{CE} changes. Transistor action such as this is the basis of the regulator circuit operation in *Figure 3-5*.

SERIES-PASS FEEDBACK REGULATOR

To gain more insight into the operation of a regulator, let's examine in detail the series-pass feedback regulator of *Figure 3-5*. The control element is NPN transistor Q_2 which is connected in series between the input voltage, V_{IN}, and the output voltage, V_O, thus, the name series-pass. The load current, I_L, is the same as I_C of Q_2, therefore, all of the load current must pass through Q_2. As we learned from the discussion of transistor action, I_C cannot flow unless there is an I_B into Q_2, thus, I_B controls I_C and I_L. How this affects the regulator operation will be discussed in a moment.

Figure 3-5 is called a feedback regulator circuit because it is a closed loop that feeds back a portion of the output voltage and compares it to a reference voltage. The difference between the two voltages determines the action that must be taken to keep the output constant.

Reference Voltage

The reference voltage, V_{REF}, in *Figure 3-5* is the voltage across diode D_1, which is a special diode called a zener diode. It is operated in the reverse-biased direction, which was described in Chapter 2 as the breakdown region. Ordi-

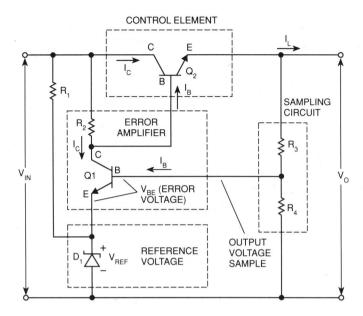

Figure 3-5. Simple Feedback Regulator

nary diodes would be damaged if operated in this region, but zener diodes are designed to be operated in this region.

In *Figure 3-5*, the input voltage to the regulator is applied to R_1 in series with D_1. Once the breakdown voltage of the zener diode is exceeded, the voltage across D_1 remains fairly constant for wide variations of current through it. The value of R_1 is chosen so the required minimum current passes through D_1 when V_{IN} is at its lowest value.

Sampling Circuit

The sampling circuit in *Figure 3-5* consists of two resistors, R_3 and R_4, in series across the output voltage terminals. The output voltage sample is the voltage across R_4. This voltage depends on the ratio of the resistor values. The values of R_3 and R_4 are chosen so the voltage across R_4 is 0.7 volt above the reference voltage across D_1. As stated previously, 0.7 volt is the V_{BE} voltage of a silicon transistor.

Error Amplifier

Q_1 is the error amplifier and the V_{BE} of Q_1 is the error voltage. The constant reference voltage is fed to the emitter of Q_1 and the output voltage sample is fed to the base of Q_1. Thus, any change in V_O varies the V_{BE} of Q_1, which changes the base current, I_B, of Q_1. Changes in I_B cause the collector current, I_C, of Q_1 to change.

Q_1 provides current gain due to transistor action. The collector current will change from 50 to 200 times the change in base current depending on the type of transistor used for Q_1. R_2 completes the collector circuit for Q_1 to V_{IN} as its supply voltage. The connection from the collector of Q_1 to the base of Q_2 provides the error-amplifier control to the control element, Q_2.

Control Element

As stated previously, Q_2 is a silicon NPN transistor. V_{IN} is designed to be larger than V_O so that Q_2 will always have enough V_{CE} for the load current range of the power supply, and still not be too large to exceed the power dissipation and temperature limits of Q_2. The power dissipation in Q_2 at any operating point is the V_{CE} across Q_2 times the current through Q_2.

The emitter of Q_2 is connected to the regulator output voltage terminal. The base of Q_2, which is connected to Q_1's collector, is at 0.7 volt above Q_2's emitter. The base-emitter circuit of Q_2 is similar to the circuit of *Figure 3-4b*. Base current is supplied to Q_2 through R_2 from the supply voltage, V_{IN}. I_B must be large enough to provide the maximum rated power supply load current ($I_C = I_L$). If Q_2 has a minimum gain of 50, then the minimum base current that must be supplied through R_2 is the rated load current divided by 50.

Q_1 controls Q_2 when the collector current of Q_1 shunts away (reduces) base current from Q_2.

Regulator Action

Now put it all together and look at the overall regulator action. The circuit is in a stable operating condition, and then a load current decrease occurs.

A decrease in load current tends to increase V_O. An increase in V_O increases the base current of Q_1, which increases the collector current of Q_1. The increased I_C of Q_1 shunts away base current from Q_2. The reduced base current of Q_2 reduces its collector current which increases its collector-to-emitter voltage. The increased voltage drop across Q_2 reduces V_O.

An increase in load current causes opposite actions. Similar regulator control loop action occurs to keep V_O constant if V_{IN} increases or decreases. The reader should go around the loop and verify the action.

IC REGULATORS

Various combinations of the sampling element, error amplifier, and control element are available in integrated circuit form. These devices make building a regulated power supply a much simpler task than it was just a few years ago. Three common voltage regulator ICs are presented in this chapter.

7800-Series Fixed-Output Voltage Regulators

The LM7800-series three-terminal regulator, as shown in *Figure 3-6*, is a series-pass regulator which has a V_{IN} terminal, a V_{OUT} terminal, and a ground terminal. Rated output voltages from 5 to 24 volts are available in the 7800 series. The last two digits of the part number indicate the output voltage of the regulator. For example, the 7805 is a 5-volt regulator, the 7812 is a 12-volt

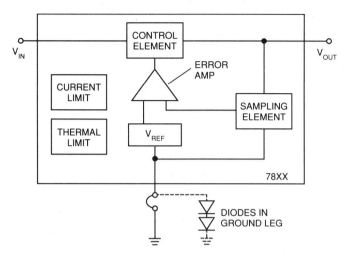

Figure 3-6. 7800-Series Regulator

regulator, and the 7815 is a 15-volt regulator. The regulated output voltage is fixed for each regulator.

The regulator contains all of the functions shown in *Figure 3-3*. In addition, it contains protection circuits to limit the peak output current to a safe value, and to maintain the internal power dissipation to a safe limit. A thermal shutdown circuit operates if the internal power dissipation exceeds a preset value. With proper heat sinks, the 7800-series can supply 1.5 amperes of load current. Notice that in order for the output voltage to be at the rated value, the sampling element must have a good ground point close to the load.

As shown in *Figure 3-6*, the output of the regulator can be increased by raising the regulator ground to a voltage above ground. The dotted lines show two 1N4001 diodes connected in series with the regulator's ground pin. Due to the forward-bias voltage drop of the silicon diodes, the ground pin of the regulator will be 1.3 to 1.4 volts above the ground level, causing the output to be 1.3 to 1.4 volts above the rated fixed-voltage output. Regulation suffers slightly because only a portion of V_O is being sampled. Higher offsets can be obtained by using more diodes, or by using a reverse-biased, high-current zener diode. Such techniques are useful to increase the fixed output voltage by a volt or two.

LM317 Adjustable Voltage Regulator
The LM317 three-terminal adjustable voltage regulator is similar to the 7800 regulator, except that it doesn't have an internal sampling element, and the ground terminal is replaced by an adjustment (ADJ) terminal. This terminal is connected to an external voltage divider which serves three purposes. The first is to provide a sampling-element voltage to the built-in error amplifier. The second is to provide enough current from ground through R1 to operate the

circuit's built-in voltage reference. The third is that it makes it easy to design an adjustable power supply. The built-in control element, an NPN transistor, is large enough to allow the LM317T to supply 1.5 amps if adequate heat-sink surface is used. The LM317T has built-in protection for maximum current and for exceeding internal power dissipation over its rated temperature range.

The operation of the LM317 is somewhat different from the 7800 series because it is what's known as a "floating regulator." It maintains its reference voltage relative to the output voltage, V_O; therefore, whenever it operates, it adjusts its operating point so that the voltage across R_1 in the R_1, R_2 voltage divider is always equal to its reference voltage, V_{REF}. As a result, as shown in *Figure 3-7*, V_O is equal to V_{REF} (1.25V in the case of the LM317) times one plus the ratio of R_2 to R_1. If R_2 is 9 times R_1, then for an LM317, V_O is 10 times 1.25V, or 12.5 volts.

One can easily see that if R_2 is an adjustable resistor, V_O can be adjusted over a wide voltage range, provided V_{IN} is within the proper range. The minimum V_{IN} must be greater than $V_O + V_{REF}$. The maximum V_{IN} cannot be greater than V_O + 40 volts; for example, if V_O=12 volts, V_{IN} cannot be greater than 52 volts. Of course, the greater V_{DIFF}, the greater the internal thermal dissipation.

Remote Sense

Another advantage of the external sampling element is that remote sensing is made easy. In many system designs, long power distribution leads cannot be avoided. A long power supply lead that has a high current will have a signifi-cant voltage drop which is outside the regulation loop. To solve the problem, remote sensing is used. A separate wire is connected to the actual load point and feeds back the load voltage directly to the sampling circuit. Because there is very small current in the wire, the voltage drop is negligible, and the regula-tion at the load point is much improved.

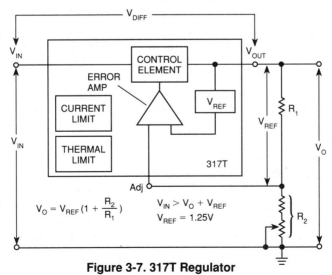

Figure 3-7. 317T Regulator

723 Regulator IC

The 723 regulator is shown in *Figure 3-8*. It contains a reference voltage source, an error amplifier, a low-current control element, and two extra components—a zener diode and a transistor. The extra transistor is used as a current limiter. As illustrated, using the 723 requires more external components to build a power supply than the 7800 and 317 regulators; however, because the basic functions are available at the external pins, the 723 is a very versatile regulator. Low-current (150 mA) stand-alone supplies use the 723, but its most common use is to drive a much larger current-handling control element.

Adding External Transistor For Higher Output Current

Figure 3-8 shows how an external transistor that can handle much larger current is connected to the 723 to provide a higher-current control element. The collector of the external transistor is connected to V_C of the 723. The base is connected to V_{OUT}, and the emitter becomes the new V_{OUT} terminal for the larger-current regulator. The 723 becomes a driver for the external control element. A proper heat sink must be used to keep the external transistor in the safe operating area.

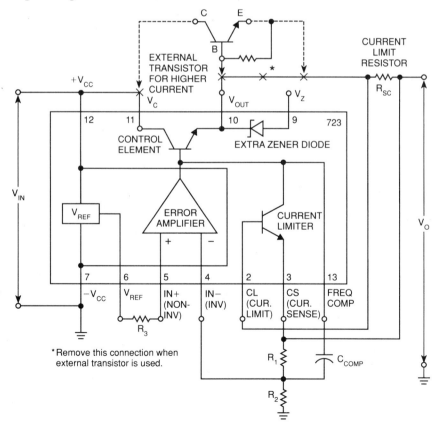

Figure 3-8. 723 Regulator

REGULATOR PROTECTION

A power supply regulator handles a considerable amount of energy while doing its job. If the electrical and thermal forces get out of control, the circuit can destroy itself. Manufacturers of IC regulators, and other IC semiconductor devices that may be added externally, specify maximum voltage, current, and temperature limits that must not be exceeded. When a circuit design stays within these limits, all components are in their safe operating area. Operation beyond these limits may destroy the device. To assure that regulators will not be destroyed if limits are exceeded, protection circuits are added to the regulators.

Short-Circuit Protection

A common occurrence is that a regulator's output terminal is shorted to ground. When this happens, the regulator tries to supply a large current beyond its rated value. As previously discussed, short-circuit protection circuits are built into 7800-series and LM317T regulators. For the 723, as shown in *Figure 3-8,* the current limiter transistor is connected to measure the voltage drop across a low resistance in series with the output terminal. As the load current increases, the voltage drop across this resistance increases. If this voltage rises above the V_{BE} of the transistor, it indicates excessive current demand by the load. The current limiter transistor conducts and shunts base current from the control element to limit its I_C, which is the load current. This keeps the control element in its safe operating area.

Thermal Runaway

Excessively high temperatures cause real problems in the operation of IC regulators or any other solid-state circuit. If the maximum junction temperature specified on a device's data sheet is exceeded, the device is usually destroyed. Junction temperature rises because of power dissipation. The expected power dissipation generated by a series-pass regulator is calculated by multiplying the voltage drop across the control element ($V_{IN} - V_{OUT}$) by the current flowing through it. This power is converted to heat which raises the junction temperature. If this heat is not removed more quickly than it builds up, a condition called thermal runaway can occur. The heat causes the device to conduct more current, resulting in more heat, until the device literally destroys itself. As long as the limits for the particular device are not exceeded, the device will remain in its safe thermal operating area.

Thermal Conduction

To protect semiconductor junctions from excessive temperatures, a path must be provided to conduct the heat away from the silicon substrate where it is generated. The usual path that the heat must travel is from the silicon to the IC package or case to the surrounding air. If the device is operated well below its maximum ratings, this usually is an adequate conduction path; however, as the power increases, the heat must be removed more rapidly.

Heat Sink

A heat sink is a piece of metal attached to a semiconductor device case or an IC package for the purpose of conducting heat away from the device. The larger the surface area of the heat sink, the faster heat is removed by conduction and radiation into the surrounding air. *Figure 3-9* shows how the metal, which is a much more efficient thermal conductor, keeps the IC from becoming too hot. Heat flows from the silicon chip, to the IC package, then to the air through the leads, printed circuit board, and heat sink.

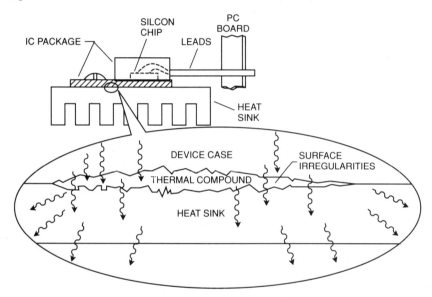

Figure 3-9. Heat Flow From IC to Heat Sink

To increase the efficiency of the heat transfer from the package to the heat sink, a thermally conductive compound is often used between them. If it is necessary to electrically insulate the device from the heat sink, a mica insulator is used. Radio Shack's power semiconductor mounting kit, 276-1371, includes such an insulator.

Sometimes the conduction and radiation cooling of the heat sink is not enough to keep the power device below its maximum junction temperature. In this situation, a fan is used to move air across the heat sink to increase convection cooling. In any design, temperature measurements should be made to ensure that maximum temperatures are not being exceeded.

SUMMARY

What has been learned? — how an available ac line voltage can be transformed to a level close to the dc voltage needed, how rectification and filtration can give a steady dc voltage, and how a regulator provides a precise voltage even under varying input voltage variations and output current demands. In Chapter 4, these principles will be put to work in useful projects.

Linear Power Supply Projects

This chapter contains three power supply projects using the principles developed in the previous three chapters. These projects use commonly available parts from Radio Shack.

REGULATED 5-VOLT POWER SUPPLY

The first project is a regulated +5V power supply that can be used to power digital circuits. This is an ideal power supply for experimentation using IC breadboards such as Radio Shack's No. 276-169.

Requirements

Digital integrated circuits are available for nearly any purpose in a wide range of complexities. As varied as they are, the vast majority require a supply voltage of 5 volts regulated to within 10%; however, some of the more complex ICs, such as microprocessors, require 5% regulation.

Regulation

The percentage of load regulation is expressed as follows:

$$\%\text{Load Reg} = (V_{NL} - V_L)/V_L \times 100$$

It is the change in voltage, expressed as a percentage, that occurs when the load changes from no-load (NL) to a load (L) divided by the load voltage. It is a measure of how far the output voltage varies in response to rapid changes in current demand by the load. Let's verify the output voltage of a 5V supply with a regulation of 5% using the load regulation equation. In the following calculation, V_v is the $V_{NL} - V_L$ variation in the above equation.

$$\pm 5\% = V_v/5 \times 100$$
$$\pm 0.05 = V_v/5$$
$$+V_v = 5(+0.05) = +0.25$$
$$-V_v = 5(-0.05) = -0.25$$

Therefore, with ±5% regulation, the 5V output voltage will vary from 5.25 to 4.75 volts regardless of the rapid peaks and dips in current demand. This regulation also applies to input voltage changes.

Load Current

The current demands of digital ICs vary from a few milliamps for logic devices to a few hundred milliamps for complex ICs. Power switching devices often control load currents of 100 mA or more which drive motors, contactors, lamps or relays. This current is in addition to the current for the logic's internal circuitry. As a result, a +5V supply with a maximum rated current of 1A (1000 mA) and a possible current of 1.5A (1500mA) should be sufficient to handle small projects and experiments that consist of around 10 ICs and some power devices.

Ripple

The ripple specification is derived using the input capacitor filter equation (from Chapter 2) and the fact that the regulator compensates for ripple voltage variations at its input. This regulator characteristic is called ripple rejection.

Digital logic circuits usually have a minimum of 200 millivolts noise margin around critical logic levels in their system design. Any power supply ripple reduces the amount of "headroom" (tolerance) on this design margin. Thus, if ripple voltage were to cause logic levels to vary by 20 millivolts, it would reduce the noise margin by 10%.

An input capacitor filter, CF_1, with a capacitance of 4700μF is chosen for this project. At a 1500mA load, the ripple voltage is:

$$V_{r(rms)} = 2.4 \times 1500/4700$$
$$V_{r(rms)} = 0.76 \text{ or } 760 \text{ millivolts}$$

The 7805 regulator will be used in the design. It has a ripple rejection characteristic of 60 decibels. The decibel, which is abbreviated dB, is the unit used to express the ripple voltage that appears at the output of the regulator when a certain voltage appears at the input. The relationship (a reduction) is expressed as follows:

$$-60\text{dB} = 20\log_{10} V_O/V_{IN}$$
$$-3 = \log_{10} V_O/V_{IN}$$
$$10^{-3} = V_O/V_{IN}$$

It means that the output ripple voltage, V_O, will be reduced by 1000 times over what appears at the input, V_{IN}. Since the input ripple voltage is 760 millivolts, it should be 0.76 millivolts at the regulator output. This is small enough to cause no problem with noise margin.

The schematic diagram for the full supply is shown in *Figure 4-1*.

Protections

Here are a number of protections built into the design.
1. The power switch, S1, is a double-pole double-throw (DPDT) switch with the line cord connected to one of the outside pairs of terminals. This isolates the line voltage inside the switch body when the switch is off.
2. An input fuse is in series with one of the primary leads of the transformer to protect against transformer shorts.
3. An LED indicator is wired across the output terminals as a pilot light and an aid in determining when the filter capacitors have discharged.
4. A fuse in series with the output helps prevent an experiment from "going up in smoke." It has been included to demonstrate its use in a design. It really

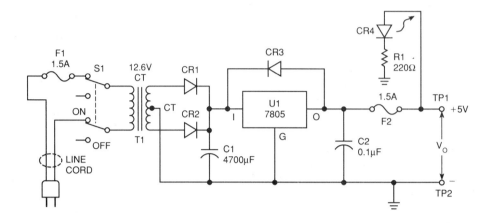

Figure 4-1. Regulated +5V Power Supply Schematic

isn't necessary because the 7805 regulator has current limiting. If the regulator used has no short-circuit protection, the fuse is absolutely necessary for experimental supplies like this.

5. Since this supply may be used with a variety of loads, a diode is connected from the input to the output so that it is reverse biased under normal operation. This provides protection if a large amount of capacitance is connected to the output. When the input voltage is turned off, the large capacitance on the output may discharge more slowly than the capacitance on the input. This would reverse bias the regulator and could damage it. The diode provides a "safe path" for the discharging current.

Regulator Choice

The requirements for this regulated +5V power supply project closely match the ratings of the fixed-voltage 7805 IC regulator—5 volts output voltage and 1.5 amperes maximum load current. The built-in thermal cutoff circuit and current limiting features are especially useful to provide additional protection when powering your experiments.

Since the 7805 IC will be delivering a power supply rated current of 1 ampere and a possible maximum current of 1.5 amperes, and since its minimum standoff voltage is 3 volts, it will dissipate a minimum of 4.5 watts at maximum load. It must be and is mounted in a heat sink and cooling holes are drilled in the enclosure to prevent heat build up.

Transformer and Rectifier Choice

The 7805 IC regulator requires a minimum voltage differential between the input and output of 3 volts. Thus, to obtain a 5-volt output, the input to the regulator must be at least 8 volts. The transformer's center-tapped secondary is rated at 12.6 volts at 3 amps. Using the two-diode full-wave rectification scheme produces 8.9 volts peak (6.3 × 1.414) at 3 amps. The rectifier diodes, which require a PIV of 17.8 volts, and the protection diode across the regulator are rated at 3A, 50V PIV. By choosing an 18V centertapped transformer (RS# 273-1515), the supply would be able to operate at 1.5A continuously because the peak voltage is raised. However, the heat sinking of the 7805 regulator has to be more extensive to protect the regulator from exceeding the power dissipation and case temperature limits.

Filters

The filtering and resultant ripple voltage is as previously described for the 7805 regulator. A small capacitor (0.1µF) is placed across the output of the regulator to bypass high-frequency noise that may be generated by the load.

Construction Details

The parts listed in *Table 4-1* are used for this project.

Table 4-1. Regulated +5V Power Supply Parts List

Description	Quantity	Reference Designator
12.6V transformer	1	T1
Diode, 1N5400	3	CR1 – CR3
IC, 7805 voltage regulator	1	U1
LED indicator	1	CR4
Capacitor, 4700µF, 35V electrolytic	1	C1
Capacitor, 0.1µF, 50V ceramic	1	C2
Resistor, 220Ω, ¼"W	1	R1
Heat sink	1	
Heat sink mounting hardware	1	
Thermal compound	1	
DPDT switch	1	S1
Circuit board (2¾" × 3¾")	1	
Box (2⅜" × 4⅜" × 7¾")	1	
Fuse holder	2	
1.5A fuse	2	F1, F2
Line cord	1	
Strain relief	1	
Binding post set	1	TP1, TP2
Rubber feet (set)	1	
Standoffs	4	
Heat shrinkable tubing		
Screws 6-32, 3/4" long		
Nuts 6-32		
Screws 4-40, 1/4" long		
Nuts 4-40		

Construct the circuit on the circuit board following the connection diagram in *Figure 4-2* and the photograph in *Figure 4-4*. *Figure 4-2* shows the underside of the circuit board. Insert the component leads through the board in the positions shown, and bend the leads as shown against the bottom of the board to hold the components in place. Besides the leads, extra lengths of bare or insulated wire are used to make the connections.

Use thermal compound between the IC and the heat sink when mounting the regulator IC. Use a 6-32 screw and nut to hold the IC in place. Insert the regulator pins through the holes in the board, wrap the component leads and wire connections around the pins, and then quickly solder the connections without heating the IC excessively. Remember, shiny solder connections are proper connections, not grainy gray. The solder on the IC leads should wet the contacts on the board to hold the IC in place, but make sure the leads do not short together.

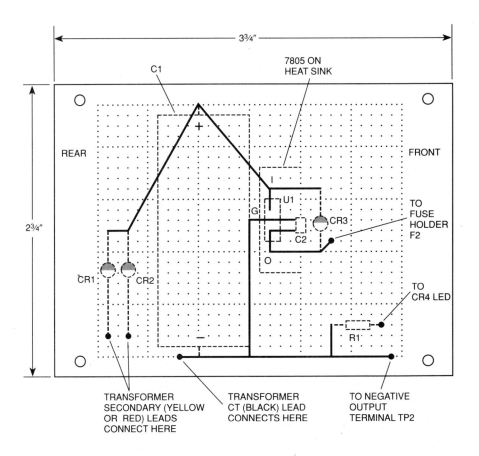

Figure 4-2. Regulated +5V P.S. Circuit Board Parts Placement (Underside shown)

Enclosure-Mounted Components

Remove the bottom cover from the box. Refer to *Figure 4-14a* for a pictorial view. Drill mounting and cooling holes in the box in the positions shown in *Figure 4-3*. The 1/8" holes are for mounting the circuit board and the transformer. The front panel holes are for fuses, binding posts, an indicator, and a power switch. The 3/8" end hole is for the line cord.

While the circuit board is unmounted, insert the two secondary (red or yellow) transformer leads through the board, and solder each to a formed diode lead on the underside of the board. Do the same with the centertap (black) lead, but solder it to the negative lead of capacitor C1. Insert a red wire through the board at the junction of CR3, C2, and the output pin of the 7805 and solder it in place. This is the positive output voltage lead to F2. Insert a black wire through the board at the negative lead of C1 and solder it in place. This is the negative output voltage lead to terminal TP2. Insert another black wire through the board and connect it to R1. This wire will be connected to the LED. Mount the standoffs to the circuit board with the short 4-40 screws.

Assembly Steps

1. Mount the transformer upside down directly to the box top in the position shown in *Figure 4-3*. Use 6-32 screws and nuts.
2. Install the fuse holders, binding posts, LED, and power switch in the front panel as shown in *Figures 4-3* and *4-4*. Mounting hardware is on the components.
3. Mount the circuit board on standoffs inside the box in the position shown in *Figure 4-3*. Use the short 4-40 screws.
4. Insert the ac line cord through the strain relief and anchor the strain relief in the end 3/8" hole. Allow 2" of slack inside the box.
5. Solder the primary leads of the transformer to the center pair of terminals on the power switch. The switch handle up indicates the power supply is on and that the bottom terminals are internally connected to the center terminals.
6. Solder one wire of the ac line cord to the side contact on the fuse holder for F1. Run a wire from the end contact of the fuse holder to one terminal of the bottom pair of terminals on the power switch. Solder the other wire of the ac line cord to the other bottom switch terminal. Insert a 1.5A fuse in the holder.
7. Solder the negative black wire from the circuit board to the lug from the black binding post TP2, then secure the lug to the post with the nut provided. This is the negative terminal of the output voltage.
8. Solder the positive red wire from the circuit board to the end terminal of the fuse holder for F2. Insert a 1.5A fuse in the holder.
9. Solder a red wire from the side terminal of the fuse holder, F2, to the red binding-post (TP1) lug. Secure the lug to the post with the nut provided. This is the positive terminal of the output voltage.
10. Solder the black wire from R1 to the short lead of the LED.
11. Slip a small piece of heat-shrinkable tubing over a red wire and solder it from the long lead of the LED to the lug on TP1. Pull the tubing up over the LED connection.
12. Recheck all connections against the schematic in *Figure 4-1*. Correct any errors.
13. Attach the cover with the self-tapping screws provided. Mark the position for the rubber feet, strip off the self-adhesive covering, and stick the feet in place.
14. Connect a voltmeter to the output, plug in the ac cord, and flip the power switch on to test the output voltage before using the supply. The open-circuit output voltage should be 5 volts and it should not change more than 5% when a resistive load drawing rated load of 1 amp (5 ohms) is placed across it.

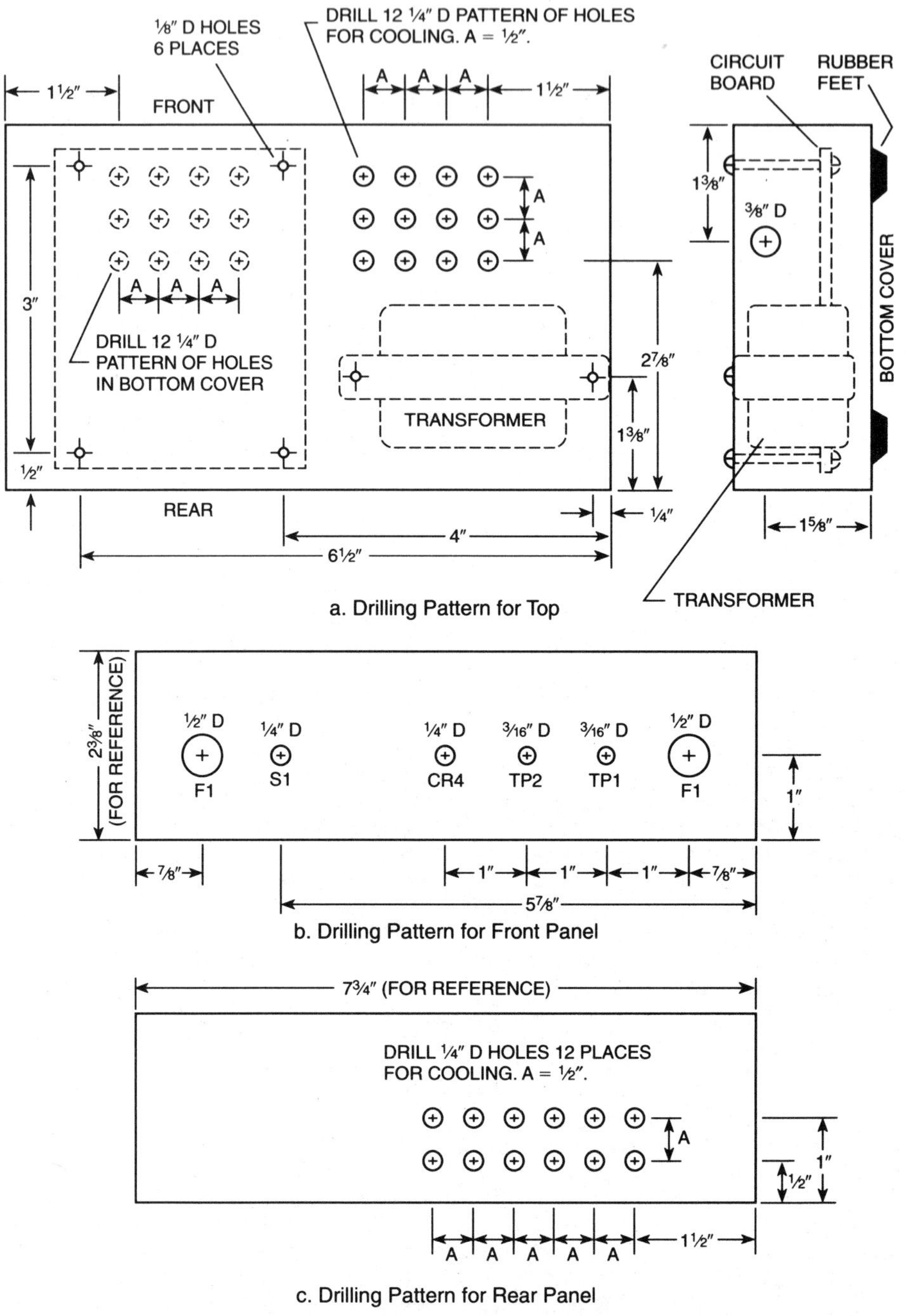

a. Drilling Pattern for Top

b. Drilling Pattern for Front Panel

c. Drilling Pattern for Rear Panel

Figure 4-3. Chassis Layout for +5V Power Supply

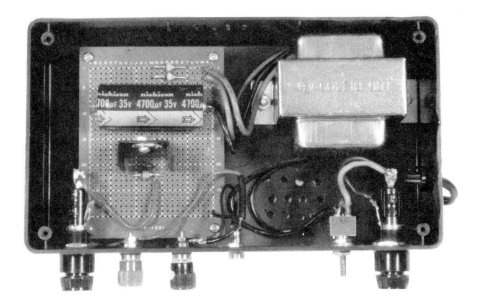

Figure 4-4. Fully Assembled +5V Power Supply

REGULATED ±12V POWER SUPPLY
This project is a dual power supply with a +12V output and a −12V output around a common ground.

Requirements
Output Voltage
Many analog ICs and some digital data communication ICs operate on differential power supplies. Such power supplies usually have a central ground with equal positive and negative voltage outputs. There are specialized ICs that require different positive and negative voltages; if this is a requirement, the design principles for setting the voltages are the same.

Load Current
Many analog applications are for small-signal amplification. As a result, the load currents are quite small; therefore, a rated load current of 1.5 amperes for each differential voltage should be quite adequate for many systems. Even if the application is a power amplifier that delivers 10 to 12 watts, the design should be adequate.

Ripple
Small-signal analog circuits require even lower ripple voltages than logic circuits; therefore, the regulator chosen for this design must have an even higher ripple rejection characteristic than the +5V supply. If the supply needs

further ripple reduction, more filter sections can be added to the input filtering, or, as will be mentioned for the LM317 regulator, additional components can be added to the regulator circuit.

Regulation
It is not likely that the current changes due to signals are going to be as severe in analog circuits as in digital circuits; however, much smaller voltage changes become significant signals in the analog circuits. For this reason, the regulation of the differential supply voltages is set at 0.5%, with a maximum of 1% over wide temperature swings.

Protection
Similar protection features to those included in the +5V supply are included in this design—a primary fuse, a DPDT power switch, and an LED pilot light.

Component Choices
Because negative regulator ICs are not as widely available as positive regulator ICs, this project uses two LM317T adjustable regulators and separate 25.2-volt center-tapped transformers for each side.

The schematic diagram in *Figure 4-5* shows two independent power supply circuits connected in series at their output, with the ac input in parallel. This is possible because of the isolation provided by the power transformers. The positive lead of the second supply is connected to the negative lead of the first, forming the central ground. The specifications of the supply are:

Output voltage: ±12V, 0.5% typical, common central ground.
Output current: a maximum of 1.5 amps for each voltage.
Ripple: 65 to 80 dB typical rejection.

The peak voltage output across the capacitor input filter is 17.8 volts (12.6 × 1.414). Two 1N5400 diodes are used for full-wave rectification, with the centertaps of the transformers used as the negative terminals. The diodes are rated at 3 amps, 50 PIV, well within the 35.6V PIV required.

The LM317T regulator offers the added advantage that the output voltage can be changed simply by changing the ratio of the resistance between the adjustment pin and ground and the resistance between the adjustment pin and the output. For 12 volts output, using 390-ohm resistors for R1 and R3, the programming resistors R2 and R4 are calculated by using the equation:

$$V_O = 1.25[1+(R2/R1)]$$
$$V_O/1.25 - 1 = R_2/R_1$$

Substituting 12V for V_O gives:

$$12/1.25 - 1 = R_2/R_1$$
$$9.6 - 1 = R_2/R_1$$
$$8.6R_1 = R_2$$

If $R_1 = R_3 = 390\Omega$, then $R_2 = R_4 = 3300\Omega$.

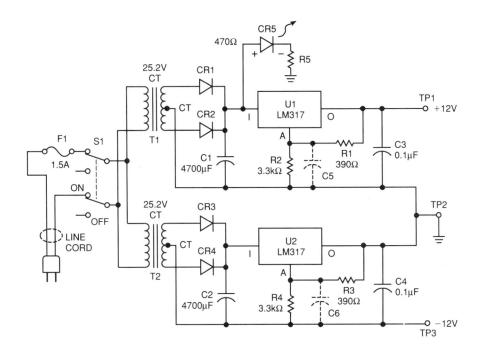

Figure 4-5. Regulated ±12V Power Supply Schematic

The LM317 has a typical ripple rejection characteristic of 65dB. It can be raised to 80dB by bypassing R_2 and R_4 with a 10μF capacitor (shown as C5 and C6). Since the same capacitive input filter is used as in the +5V supply, the input ripple will be the same—760 millivolts. With 65dB rejection, the output ripple is over 2000 times less. At 80dB rejection, the ripple at the output is reduced by 10,000 times.

Construction Details

The parts listed in *Table 4-2* are used for this project.

Construct the circuit on the perfboard following the connection diagram in *Figure 4-6*. Scribe the perfboard along a row of holes to obtain the 3″ × 6″ dimensions, then snap the board along the scribe line. As in the other projects, insert the leads of the components through the board in the position shown, bend the leads over against the bottom of the board, and make the connection to the underside of the board as shown in *Figure 4-6*. Add extra wire to complete all connections, and solder component leads and wire together.

Insert through the board, at the positions shown in *Figure 4-6*, two red wires, one black wire, and two wires of another color. Solder one red wire to the negative output, TP3, and the other red wire to the positive output, TP1. Solder

Table 4-2. Regulated ±12V Power Supply Parts List

Description	Quantity	Reference Designator
25.2V transformer	2	T1, T2
Diode, 1N5400	4	CR1 – CR4
IC, LM317 voltage regulator	2	U1, U2
LED indicator	1	CR5
Capacitor, 4700µF, 35V electrolytic	2	C1, C2
Capacitor, 0.1µF, 50V ceramic	2	C3, C4
Resistor, 390Ω, 1/2W	2	R1, R3
Resistor, 3.3kΩ, 1/4W	2	R2, R4
Heat sink	2	
Heat sink mounting hardware	2	
DPDT switch	1	S1
Box (3¹⁄₁₆″ × 8¼″ × 6⅛″)	1	
Perfboard (4½″ × 6″)	1	
Fuse holder	1	
1.5A fuse	1	F1
Line cord	1	
Strain relief	1	
Binding post set	1	TP1
Rubber feet (set)	1	
Standoffs	4	
Screws 6-32, 3/4″ long		
Nuts 6-32		
Screws 4-40, 1/4″ long		

the black wire to the ground connection, TP2. The other colored wires will connect to CR5. Solder one to the junction of CR1 and CR2, and the other to the end of R5. Mount the standoffs at each corner of the board with short 4-40 screws.

Enclosure-Mounted Components
Drill mounting and cooling holes in the bottom and top of the enclosure as shown in *Figure 4-7*. Refer to *Figure 4-14b* for a pictorial view. The transformer and perfboard mount to the inside of the bottom. Install the rubber feet to the bottom with the self-adhesive pads in the position indicated in *Figure 4-7*.

Assembly Steps
1. Install the binding posts and LED to the front panel portion of the bottom, and the power switch and fuse holder to the rear panel as shown in *Figures 4-7* and *4-8*.
2. Insert the ac line cord through the strain relief and anchor the strain relief in the 3/8″ hole in the rear panel. Allow 2″ of slack inside the bottom.
3. Mount the transformers to the bottom near the rear panel as shown in *Figure 4-7*. Use 6-32 screws and nuts.
4. Solder one wire of the ac line cord to the end contact on the fuse holder or F1. Run a wire from the fuse holder's side contact to one terminal of the bottom pair of terminals on the power switch, S1. Solder the other ac line cord wire to the other bottom switch terminal. Solder the primary leads of the transformers to the center pair of terminals. Insert a 1.5A fuse in the fuse holder. When the switch handle is up, the power supply is on.

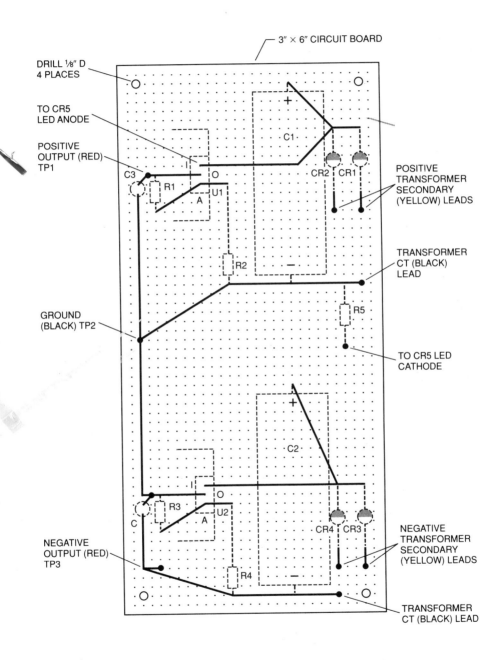

3″ × 6″ CIRCUIT BOARD

DRILL ⅛″ D
4 PLACES

TO CR5
LED ANODE

POSITIVE
OUTPUT (RED)
TP1

C3

R1

U1

O

A

C1

+

CR2   CR1

POSITIVE
TRANSFORMER
SECONDARY
(YELLOW) LEADS

R2

−

TRANSFORMER
CT (BLACK)
LEAD

GROUND
(BLACK) TP2

R5

TO CR5 LED
CATHODE

C2

+

NEGATIVE
OUTPUT (RED)
TP3

R3

U2

O

A

C

CR4   CR3

NEGATIVE
TRANSFORMER
SECONDARY
(YELLOW) LEADS

R4

−

TRANSFORMER
CT (BLACK) LEAD

**Figure 4-6. Regulated ± 12V P.S. Circuit Board Parts Placement
(Underside Shown)**

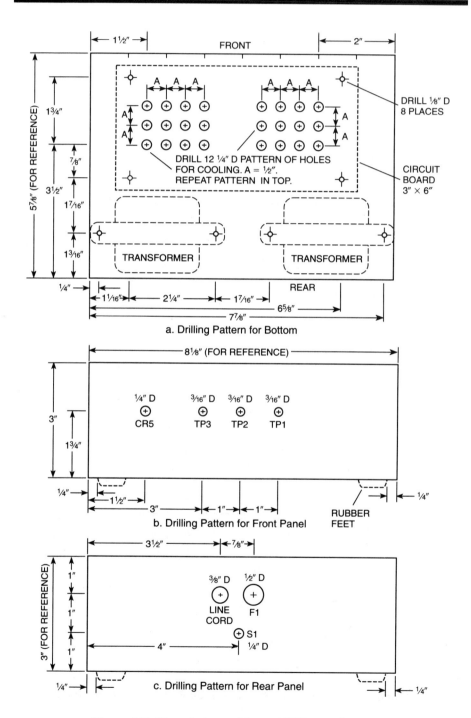

a. Drilling Pattern for Bottom

b. Drilling Pattern for Front Panel

c. Drilling Pattern for Rear Panel

Figure 4-7. Chassis Layout for ± 12V Power Supply

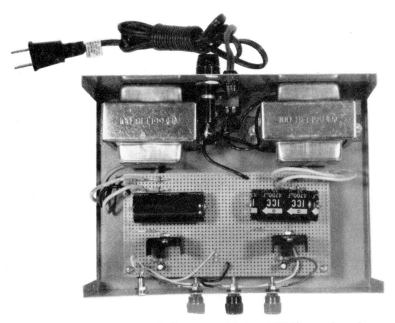

Figure 4-8. Fully Assembled ±12V Power Supply

5. While the perfboard is unmounted, insert the two secondary (yellow) leads of each transformer through the board and solder each to a formed rectifier diode lead on the underside of the board. Follow the diagram carefully. The negative transformer leads go to CR3 and CR4; the positive transformer leads to CR1 and CR2. Insert the black centertap transformer leads through the board. The negative transformer black lead is soldered to the negative output connection; the positive transformer black lead is soldered to ground. Mount the standoffs to the board with short 4-40 screws.
6. Mount the perfboard with its standoffs to the bottom of the box with short 4-40 screws into the standoffs.
7. Solder the red negative-output lead from the circuit board to the lug from the red negative-output binding post (TP3), then secure the lug to the post with the nut.
8. Solder the red positive-output lead from the circuit board to the lug from the red positive-output binding post (TP1), then secure the lug to the post with the nut.
9. Solder the black ground lead from the circuit board to the lug from the black ground binding post (TP2), then secure the lug to the post with the nut.
10. Slip a small piece of heat-shrinkable tubing over the wire from R5 and solder it to the short lead on CR5. Do the same for the remaining wire from the board, but solder it to the long lead on CR5. The LED indicator is now connected. Slip the tubing over the connections to CR5.
11. Recheck all connections against the schematic in *Figure 4-5*. Correct any errors.
12. If cooling holes have not been drilled in the top, do that now.
13. Connect a voltmeter to the output, plug in the ac cord, and flip the power switch on to test the output voltages before using the supply. Each output voltage should be 12 volts and remain within 1.0% regulation when a load from milliamperes to 1.5 ampere is placed across the supply (an 8Ω, 20W power resistor must be used for 1.5A).

CAUTION

If the supply is going to be used continuously at 1.5A, then it is best to have a fan blowing down on the top to keep the regulator within its limits of power dissipation at a particular case temperature.

VARIABLE POWER SUPPLY

The last project for this chapter is a power supply with a variable output voltage suitable for a wide range of uses. It covers approximately 3.0 – 30 volts at up to 1.5 amps of current in three ranges. The schematic is shown in *Figure 4-9.*

Transformer and Regulator

The transformer is a 25.2V, 2A, center-tapped (C.T.) transformer used with a full-wave bridge rectifier rated at 1.5A, 50V PIV, to provide 35.6 V_{pk} (minus the bridge diode drops), or a few volts higher if the ac line voltage is over 120 volts. The C.T. connection is not used. It is taped and tucked out of the way so it will not short out. The regulator is an LM317. Since the LM317 regulator is used, the typical regulation specification is 0.5% just like the previous supply. The sampling element includes a variable resistor (5KΩ) to provide a front-panel voltage adjustment.

When the output current is 1.5A, the input capacitive filter peak voltage has dropped to 27 – 28 volts. If the output voltage were set to 3.0 volts, the regulator would have to drop over 25 volts. Thus, at 1.5 amps, the regulator would dissipate over 37.5 watts as heat, which would quickly trigger the device's thermal-shutdown protection circuit.

To keep the power dissipation of the regulator within acceptable values, the operating range of the supply is divided into three ranges. Within each range, switching is designed to limit the minimum output voltage, and the input voltage to the regulator at maximum current. A voltage dropping resistor is placed in series with the regulator to control the input voltage, while the minimum voltage for each range is set by the resistor ratio in the sampling element. For example, for the low range, R1 plus R2 is 740 ohms and R6 plus R7 is a minimum of 740 ohms. The ratio of R6 plus R7 to R1 plus R2 is 1. Using the equation:

$$V_O = 1.25(1 + (R6+R7)/(R1+R2))$$

the minimum voltage for the range is 2.5 volts (1.25(1+1)). The minimum differential voltage across the LM317 is specified to be 3 volts; however, many of the regulators will operate down to 2.5 volts. By inserting a 9Ω resistor in series right after the capacitive input filter, the input voltage to the regulator is limited to 14.5 volts when the load current is 1.5 amperes. The resistor not only serves as a series dropping resistor, but also as the resistive part of another resistive-capacitive filter section formed with C2.

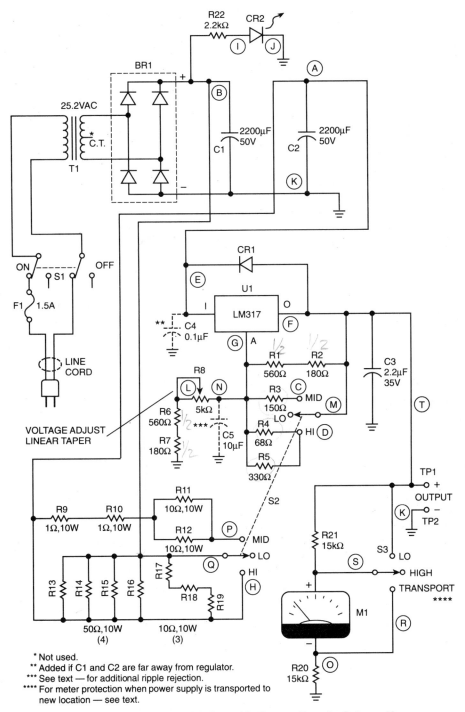

* Not used.
** Added if C1 and C2 are far away from regulator.
*** See text — for additional ripple rejection.
**** For meter protection when power supply is transported to
   new location — see text.

Figure 4-9. Regulated Adjustable Power Supply Schematic

The approximate voltage drops across the series resistance for the low and mid ranges is as follows:

Range	R_S	1.0A	1.5A
LO	9Ω	9V	13.5V
MID	4Ω	4V	6V

Ripple

As in the previous supply using the LM317, the ripple rejection is either 65dB, or, if bypass capacitor C5 is used, it increases to 80dB. Using the ripple voltage equation, the ripple voltage input to the regulator is calculated as:

$$V_{rl(rms)} = 2.4 \times 1500/2200 = 1.64 \text{ volts}$$

With 65dB of ripple rejection, the ripple voltage at the regulator output is reduced over 2000 times to below 0.82 millivolts. Actual measurements show this to be from 3 to 10 millivolts. Long leads due to the switching seem to contribute to larger ripple.

The output capacitor C3 is a 2.2μF tantalum capacitor whose main task is to prevent noise generated by the load from affecting the regulator operation. If a layout of a power supply design were to place the filter capacitors a fair distance from the regulator, then the 0.1μF capacitor C4 (indicated in dotted lines) would be added to the design.

Indicators

A LED pilot light and a meter are used to indicate voltage presence. The LED, with a current-limiting resistor (2.2kΩ), indicates the presence of dc (prior to regulation). The meter indicates the output voltage and a range switch changes the full scale from 15 volts to 30 volts. The TRANSPORT position of the switch, which shorts across the meter, should be used when the supply is being moved.

Heat Sinks

Because of the high power dissipation in this supply at the current extremes, the regulator and TO-220 heat sink are mounted to the back of the enclosure. The regulator is insulated from the heat sink. This provides maximum heat sink area for thermal protection. However, if the supply is operated continuously at voltage and current conditions near the maximum boundaries indicated in *Figure 4-10,* a fan should be directed to blow over the back of the enclosure as an added precaution to protect the regulator.

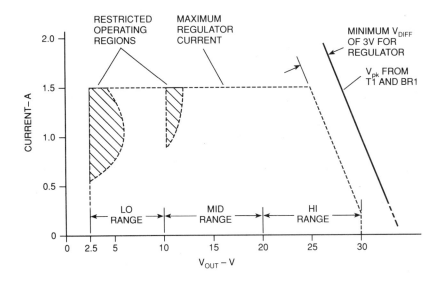

Figure 4-10. Operating Regions for Adjustable Power Supply

Operating Region

As shown in *Figure 4-10,* there are regions within the operating voltage ranges where the maximum power dissipation of the LM317 (15W @ T_C=25-90°C) is being exceeded. *The power supply should not be operated within these regions.* Other operating restrictions also are shown:

1. The peak voltage output from the transformer and rectifier at various currents and the 3V minimum differential across the LM317 limit the maximum output voltage to 25 volts at 1.5A.
2. The maximum load current for the LM317 is 1.5A.
3. The three nominal voltage ranges are:

    LO: 3.0 to 11 volts
    MID: 9.5 to 20 volts
    HI: 18.0 to 30 volts

The voltage adjustment varies with the ranges. A full turn of the knob is required on the LO range, approximately half of the sweep is required on the MID range, and less than half of the sweep is required for the HI range. As the output voltage on each range is increased, the differential voltage across the LM317 decreases, and (especially at high currents) may drop below 3 volts. This causes the LM317 to drop out of regulation.

Construction Details

The parts listed in *Table 4-3* are used for this project:

Table 4-3. Variable Power Supply Parts List

Description	Quantity	Reference Designator
25.2V transformer	1	T1
Bridge rectifier	1	BR1
Diode, 1N5400	1	CR1
LED indicator	1	CR2
LM317	1	U1
Capacitor, 2200μF, 50V electrolytic	2	C1, C2
Capacitor, 2.2μF, 35V electrolytic	1	C3
Resistor, 560Ω, 1/2W	2	R1, R6
Resistor, 180Ω, 1/2W	2	R2, R7
Resistor, 150Ω, 1/4W	1	R3
Resistor, 68Ω, 1/4W	1	R4
Resistor, 330Ω, 1/4W	1	R5
Potentiometer, 5kΩ	1	R8
Resistor, 1Ω, 10W	2	R9, R10
Resistor, 10Ω, 10W	5	R11, R12, R17 - R19
Resistor, 50Ω, 10W	4	R13 - R16
Resistor, 15kΩ, 1/4W	2	R20, R21
Resistor, 2.2kΩ, 1W	1	R22
Circuit Board (2¾" X 3¾")	1	
Heat sink	1	
Heat sink mounting hardware	1	
DPDT switch	1	S1
DPDT switch, center off	2	S2, S3
Box (3¹/₁₆" X 8¼" X 6⅛")	1	
Fuse holder	1	
Fuse, 1.5A	1	F1
Binding post set	1	TP1, TP2
Rubber feet (set)	1	
Line cord	1	
Strain relief	1	
Panel meter	1	M1
Knob	1	
Standoffs	4	
Heat Shrinkable Tubing		
Screws 6-32, 3/4" long		
Nuts 6-32		
Screws 4-40, 1/4" long		
Hookup wire #20 AWG		

Circuit Board

As with the previous projects, insert the components through the circuit board in the positions shown in *Figure 4-11*, bend the leads, and interconnect the circuit. Use extra wire (#20 AWG) as necessary. Insert three 2" wires through the board at points E, F, and G, respectively, as shown, and solder in place. Remember *Figure 4-11* is the underside of the board. Slip short lengths of shrinkable tubing over the wires. These wires connect to the LM317, which is mounted on the rear panel of the enclosure.

Figure 4-5. Regulated ±12V Power Supply Schematic

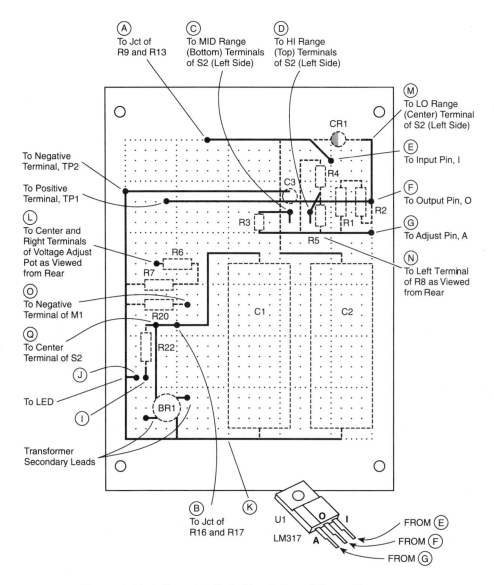

**Figure 4-11. Adjustable P.S. Circuit Board Parts Placement
(Underside Shown)**

Insert 7″ lengths of wire for A, B, C, D, I, J, L, M, N, O, and Q, and solder in place. Make these wires different colors so they can be identified easily. A will be in the main output supply line to the LM317 and connects to C2. B is the V_{pk} voltage line to the series dropping resistors that control the voltage ranges. Q is the connection from the junction of R16 and R17 to the range switch, S2, center terminal. C, D, and M connect to one side of the range switch, S2; I and J connect to CR2; L and N connect to the adjustment control, R8; and O connects to M1 for full scale of 15 volts.

Insert a 6″ red and black wire through the board at TP1 and TP2, and solder in place. TP1 will connect to the positive output voltage terminal and TP2 to the negative output voltage terminal.

Drill holes in the enclosure as shown in *Figure 4-12*. Refer to *Figure 4-14b* for a pictorial view of the enclosure. Install the rubber feet to the bottom with the self-adhesive pads.

Assembly Steps

1. Proceed carefully one step at a time. The interconnections are a bit complicated. Refer to *Figures 4-9, 4-11, 4-12,* and *4-13* at each step.
2. Install the power switch, fuse holder, and ac line cord in strain relief in the rear panel. Leave about 2″ line cord slack inside the box.
3. Solder one wire of the line cord to the end contact on the fuse holder for F1. Run a wire from the fuse holder's side contact to one bottom terminal of the power switch, S1. Solder the other line cord wire to the other bottom switch terminal. Insert a 1.5A fuse in the fuse holder.
4. Mount the transformer to the bottom of the enclosure. Solder the primary leads to the center pair of terminals of S1. When the switch handle is up, the power supply is on.
5. The power resistors are stacked and held in contact with the enclosure bottom by two wire ties, as shown in *Figures 4-12* and *4-13*. Pass the ties through the holes in the bottom of the enclosure to hold the resistors in place. Interconnect and solder the leads of R9 through R18 shown in *Figure 4-9*. Attach two wires to form the connection H and the connection P from the resistors to the range switch, S2.
6. Install the front-panel components, S2, S3, CR2, TP1, TP2, R8 and M1.
7. Mount the regulator IC on the heat sink with the insulator provided in 276-1373 between the IC and the heat sink. With the same screw and nut, attach the heat sink to the rear of the enclosure. Use thermal compound on all surfaces.
8. Before mounting the circuit board, insert the transformer secondary leads through the circuit board as indicated in *Figure 4-11,* and solder in place. C.T. is not used.
9. Mount the circuit board to the standoffs, and mount the assembly to the bottom in the position shown in *Figures 4-12* and *4-13*. Use 4-40 screws.
10. Solder wires E, F, and G to the LM317 as shown in *Figure 4-11*. Trim wires to length and slip shrinkable tubes over connections.
11. Looking at the voltage adjustment resistor, R8, from the rear, solder N to the left hand terminal. Solder L to the center terminal and right hand terminal, shorting the terminals together.
12. Slip a small length of heat shrinkable tubing over the LED wires, I and J, and solder them to the LED indicator leads. J connects to the short lead (cathode); I connects to the long lead (anode). Slip tubing over connections.
13. Solder the positive red wire to the lug of TP1, and the negative black wire to the lug of TP2. These are the output voltage binding posts.
14. Solder wire O to the negative terminal of the meter, M1. Solder a wire, R, from the negative terminal of M1 to the bottom right-side terminal (looking from rear) of S3.
15. Run a wire S from the positive terminal of M1 to the center right-side terminal of S3. Also, between this same terminal and the top terminal of S3, connect R21. R21 makes M1 into a 30V full-scale meter. Complete the M1 connections by running wire T from the top terminal of S3 to the lug of TP1. Solder all connections.
16. Solder wire A to the junction of power resistors R9 and R13. At this junction is wire H from step 4. Solder H to the top right-side (looking from rear) terminal of S2.

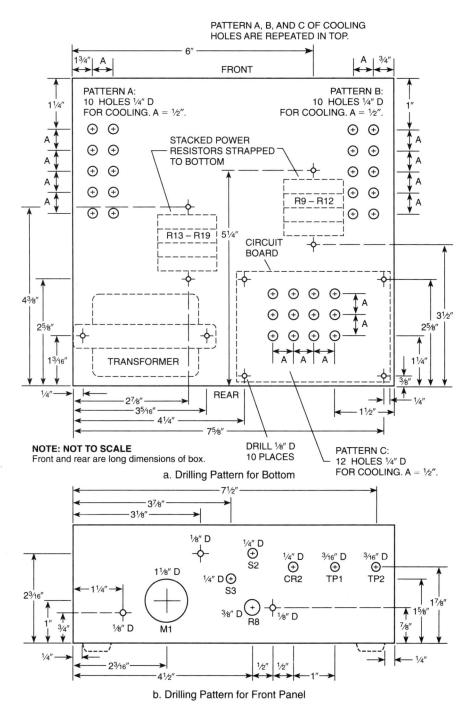

a. Drilling Pattern for Bottom

b. Drilling Pattern for Front Panel

Figure 4-12. Chassis Layout for Adjustable Power Supply

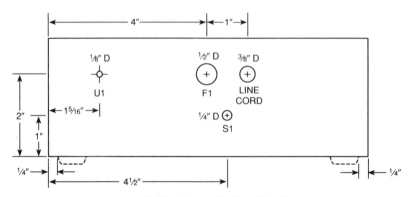

c. Drilling Pattern for Rear Panel

Figure 4-12. Chassis Layout for Adjustable Power Supply *(Continued)*

Figure 4-13. Fully Assembled Adjustable Power Supply

17. Solder wire Q to the center right-side terminal of S2, and wire P to the bottom right-side terminal. When the handle of S2 is up, S2 is on MID range; when it is down, it is on the HI range; and when it is in the center, it is on the LO range.
18. Solder wire C to the bottom left-side (looking from rear) terminal of S2, wire M to the center terminal, and wire D to the top terminal.
19. Recheck all connections against the schematic in *Figure 4-9*. Correct any errors.
20. If cooling holes have not been drilled in the top, do it now. Attach the top as a cover.
21. Connect a voltmeter to the output and test the supply using various values of power resistors. *Table 4-4* lists various voltage and power resistor combinations to test the supply at various load currents. Refer again to *Figure 4-10* to make sure the power supply is operating only in the permissible region. The regulation should be typically 0.5% over the operating region.

a. +5V Power Supply   b. ±12V and Adjustable Power Supply

Figure 4-14. Enclosures

Table 4-4. Resistance Values to Test the Variable Power Supply

Range	V_o V	I_L A	R_L OHMS	P_D* W
HI	25	1.25	20	5
	20	1.5	13.3	12
MID	20	1.5	13.3	3
	15	1.5	10	10.5
	12	1	12	14
LO	10	1.5	6.7	6.8
	6	1	6	15
	3	0.5	6	12.3

* Because of line voltage variations, the P_D values may be different. If P_D becomes too high, the thermal protection circuit may shut down the regulator.

SUMMARY

This chapter has given the design and construction details of three different power supplies that are adequate for a variety of uses. The circuits may be changed or modified as needed by the person building them.

The discussion so far has focused on linear power supplies. In the next two chapters, the basic principles of switching power supplies will be investigated.

CHAPTER 5

Switching Power Supply Systems

This chapter describes the operation of a switching power supply system which uses switching regulators. A switching regulator has much higher efficiency than the linear systems described in the first four chapters.

Linear series-pass regulators convert an input voltage that is higher than needed to a desired lower voltage. The extra energy (the voltage drop across the control element times the current flowing through it) is dissipated as heat. As a result, typical series-pass regulators have a conversion efficiency (P_{OUT}/P_{IN}) of 50% or less.

Switching regulators, on the other hand, can have a conversion efficiency of 85% or more. Such efficiency results in lower power dissipation and much smaller size components for a given power output. Other advantages are: 1. operation over a wide range of current and voltage, 2. switching-mode operation for the control element, 3. input voltage can be lower than the output voltage, and 4. the output voltage can be of opposite polarity than the input voltage.

SWITCHING POWER SUPPLY OPERATING PRINCIPLES

Figure 5-1 is a block diagram of a switching regulator power supply. As one can see, there are many similarities between switching systems and the linear systems discussed in Chapter 3. The differences lie in the action of an inductor used for temporary energy storage, and how the control element is controlled to provide regulation.

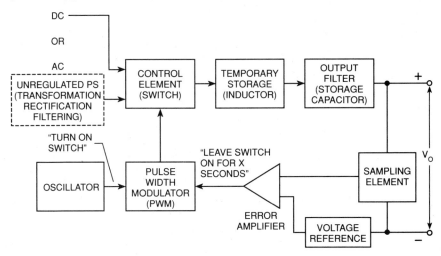

Figure 5-1. Switching Power Supply Block Diagram

If an ac source is used, the transformation, rectification, and filtering circuits that provide a dc input voltage to the regulator serve the same function in a switching power supply as in a linear power supply. If a dc source is used, an input filter may be required for ripple or noise reduction or for stability.

Control Element Action

In linear systems, regulation is accomplished by varying the resistance of the control element. In switching systems, it is done by rapidly turning the control element on and off, and by varying the ratio of ON time to OFF time. Unlike the series-pass control element, there is no linear operating state; the control element is either completely on or completely off.

The ON switching action "pumps" energy in sudden bursts into the inductor temporary storage element. During the time that the switch is OFF, the stored energy is directed by a diode into the filter capacitor to supply the load as needed. The sampling element, reference-voltage source, and error amplifier work in an identical manner to those in a linear supply. However, the output of the error amplifier is used differently.

Error Amplifier, Oscillator and PWM

The new circuits for the switching regulator are the oscillator, the pulse-width modulator (PWM), and the temporary storage element inductor. The control element is still a transistor or power field-effect transistor, but it is operated as a switch. It is turned on and off by the PWM as shown in *Figure 5-2.*

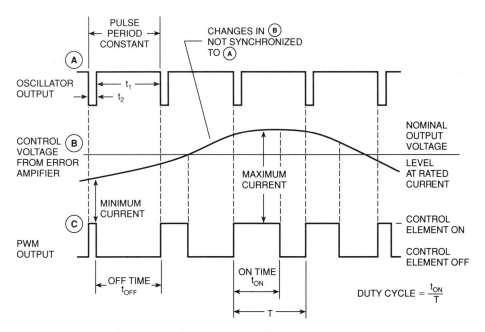

Figure 5-2. Switching and Control Waveforms

The oscillator provides output pulses at a constant frequency, and the PWM output pulses have varying ON times compared to the total period, T, of the oscillator pulses. The oscillator pulse to the PWM input tells the PWM output pulse to turn the control element ON. The error amplifier voltage level tells the PWM how long the output pulse should keep the control element ON. Thus, the error amplifier output controls the width of the PWM pulse, which controls the ON time of the control element. The duty cycle of the control element is t_{ON}/T.

Inductor Action — Storing Energy

The switching regulator cannot be understood unless the action of the temporary storage inductor is understood. *Figure 5-3a* shows an inductor in series with resistor, R, which represents the load, and switch, S1, across a battery with voltage V_{IN}. The initial conditions are: S1 is open, current is zero (I=0), and voltage across the load is zero ($V_R=0$). The inductor, L, is assumed to have no resistance.

The curve shown alongside the circuit plots the value of V_R against time. Since $V_R=IR$ and R is a constant, the curve also represents the value of I in the circuit at any instant in time. After S1 is closed a relatively long time, notice that

$$V_R = V_{IN} \text{ and } I = V_{IN}/R.$$

At the instant that S1 is closed, the current in the circuit tries to increase to the V_{IN}/R value, but the inductor action resists current change. It does this by developing a counter voltage, V_L, across itself that is in a direction to resist the current change. The counter voltage is expressed by engineers as the counter electromotive force (CEMF). (This is the same "voltage by induction" that was discussed in Chapter 1 for generators and transformers.) As the current changes, the magnetic flux extending out from the inductor as a result of the current change, cuts the turns of the inductor coil and induces the counter voltage, V_L, in the inductor. The V_L (or CEMF) developed across the inductor can be expressed by the equation:

$$V_L = L\frac{\Delta I}{\Delta t}$$

where: L = the inductance in henries
ΔI = change in current in amperes
Δt = time period in seconds of current change

At the instant that S1 is closed, the current is trying to change from zero to maximum, therefore, V_L is maximum. Since all the input voltage appears across the inductor, $V_R=0$. As time passes, the current increases in a logarithmic curve as shown for V_R, and finally reaches its maximum value, V_{IN}/R. When the current reaches maximum, it no longer changes, and the magnetic flux stops changing; therefore, $V_L=0$. Since all of the input voltage, V_{IN}, appears across R, V_R is maximum. Thus, the action of the inductor is to resist a current change and store energy in the magnetic flux field built up around it by the current through it. A similar opposition to current change that occurs when S1 is opened is described in the following paragraphs.

Inductor Action — Releasing Energy

Figure 5-3b is the same circuit as *Figure 5-3a* except diode, D1, has been added. The diode's purpose will be explained shortly. The current in the circuit is maximum, V_{IN}/R, and $V_R=V_{IN}$. Energy has been stored in the magnetic flux surrounding the inductor. Now, S1 is opened.

The curve alongside the circuit plots V_R against time. At the instant S1 opens, the current, I, wants to change to zero because S1 is open. Since I wants to change to zero, the magnetic flux field built up in L collapses. As the magnetic flux cuts the turns of wire in the inductor, the counter voltage, V_L, is again induced, but this time the polarity of V_L is opposite to what it was in *Figure 5-3a*.

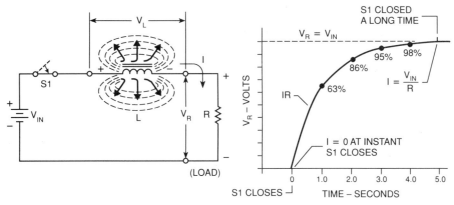

a. Action of Inductor When S1 Closes

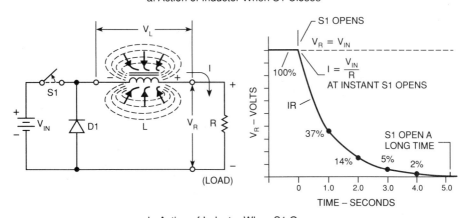

b. Action of Inductor When S1 Opens

Figure 5-3. Switched Inductor Action

The polarity is such that it wants to keep the current in the same direction as it was before S1 opened. D1 provides a complete path for I through R and L as the magnetic field collapses. (If D1 were not present, a very high voltage would be developed across the opened S1 contacts, causing the contacts to arc over to release the stored energy in L.)

Thus, the energy stored in the magnetic field of L is returned to the circuit by inducing a voltage in L to resist the change in I, and to keep I through R in the same direction it was before the current change occurred. V_R, and thus I, reduces along a logarithmic curve as the field collapses until I and V_R are zero. V_L is calculated from the same equation as for *Figure 5-3a*. The inductor action described in *Figures 5-3a* and *5-3b* is used in several different types of switching regulators.

STEP-DOWN REGULATOR

Figure 5-4a shows a control element, inductor and output filter for a *step-down* switching regulator. It is used when the required regulated voltage is lower than the input voltage. When the control element is switched on, the inductor stores energy, helps supply load current, and supplies current to the capacitor. When the control element is switched off, the energy stored in L helps supply the load current, but also again restores the charge on C_F that is supplied to the load during the time that the control element is off and L has discharged its energy. In this circuit $V_L = V_{IN} - V_O$ when the control element turns on, and $V_L = V_O$ when the control element turns off.

Alongside the circuits in *Figure 5-4*, waveforms of inductor and/or capacitor currents in the circuits are plotted against the time that the control element is on and off. *Figures 5-4a* and *5-4b* show curves of the current, I_C, into the output filter capacitor. There are times when L is supplying current to charge the capacitor ($+I_C$) and supplying the load current as well; and there are times when the load current is supplied only by the capacitor ($-I_C$). The switching regulator, when working properly, is just at balance in a switching cycle—it is supplying the capacitor as much charge, $+Q$, as the load discharges, $-Q$, from the capacitor.

STEP-UP REGULATOR

Figure 5-4b shows the same circuit parts for a *step-up* switching regulator. It is used when the required regulated voltage is higher than the input voltage. It operates slightly differently than *Figure 5-4a*. Energy is stored in L when the control element is on. The energy is supplied by V_{IN} and $V_L = V_{IN}$ during this time. The load, isolated by D1, is supplied by the charge stored in C_F. When the control element turns off, the stored energy in L is *added* to the input voltage, and I_L helps supply the load current and restores the energy discharged from C_F. C_F supplies current to the load after L discharges. When the control element turns off, $V_L = V_O - V_{IN}$.

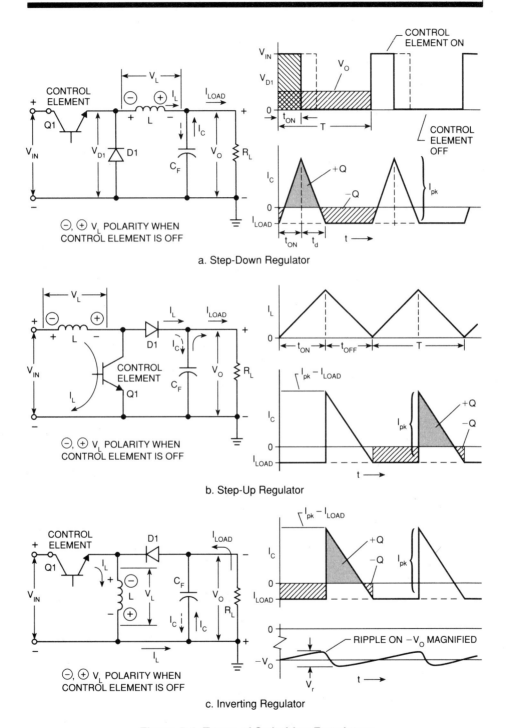

a. Step-Down Regulator

b. Step-Up Regulator

c. Inverting Regulator

Figure 5-4. Types of Switching Regulators

INVERTER REGULATOR

Figure 5-4c shows the same circuits parts for an *inverter* regulator, sometimes called a *flyback* regulator. The inverter regulator is used when a regulated output voltage of the opposite polarity of V_{IN} is required. Its operation is similar to the step-up regulator. When the control element is on, energy is stored in L, and D1 isolates L from the load. The load current is supplied by the charge on C_F. When the control element turns off, the stored energy in L charges C_F to a polarity such that V_O is negative. I_L supplies load current and restores the charge on C_F during the time it is discharging its energy. As with the step-up regulator, C_F supplies the load current after L discharges. When the control element is on, $V_L = V_{IN}$; when the control turns off, $V_L = V_O$. Depending on how the inverter is designed, it can be used for voltages higher or lower than the input.

COMPLETE SWITCHING REGULATOR

Figure 5-5 shows a functional block diagram of a complete *step-down* switching power supply. Each of the individual functions will be discussed in more detail.

Input Voltage

The input voltage for a switching power supply can vary over a surprisingly wide range while still maintaining good conversion efficiency. An ac source is indicated for V_{IN}. All the same design considerations apply for the transformation, rectification and filtering as for the unregulated supply of Chapter 2. The ripple voltage filtering, which takes care of any 60Hz or 120Hz noise, can be somewhat less because of the output filtering of the switching regulator.

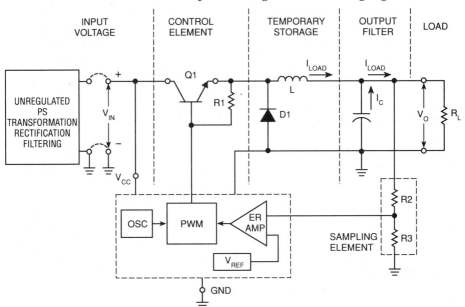

Figure 5-5. Complete Switching Regulator (Step-Down)

Control Element Switch

The control element switch is a switching power semiconductor. It must have low V_{ON} across it, and must have fast switching times. It may be a single NPN transistor or power FET, or it may be a combination NPN and PNP for higher gain operation. Regardless of what type of transistor is used, it is turned completely on (lowest resistance) during the ON time, then abruptly turned completely off (highest resistance) during the OFF time. It is used as if it were a relay, but operates at a very fast rate—from 20kHz to 100kHz. A 100 kHz wave has a period of 10 microseconds (0.00001 second).

The Catch Diode — Directing the Energy

When the magnetic field of the inductor begins to collapse and release its stored energy, the energy must be contained and channeled in a useful direction. A diode, called the *catch diode,* does this in each of the circuits of *Figure 5-4.* It directs the stored energy into the output filter capacitor. The one-way conduction of the diode is used to provide the proper circuit connection when the induced voltage across the inductor is the proper polarity. Due to the high switching frequencies in these supplies, the diode must have a very low forward voltage and a very fast switching time and recovery time. A Schottky diode is an ideal diode for the application.

The Inductor

The action of the inductor has been discussed thoroughly. The main concern is that the chosen inductor has the proper inductance, that it not saturate during its operation, and that the core has the volume to handle the power required. If it saturates, it loses its inductance and its ability to efficiently transfer energy to the output.

The inductance is determined by the core material and the number of turns of wire on the core. Ferrite and powdered iron cores are usually used for switching power supply inductors. When iron laminated cores are used, they have core loss at higher frequencies which causes the supply efficiency to be low.

The inductance of a coil can be determined from the following equation:

$$L = \frac{N^2 A u u_o}{1}$$

where: L = inductance in henries
N = number of turns
A = cross sectional area of coil (m^2)
u = permeability of core material
u_o = absolute permeability of air (1.26×10^{-8} H/m)
1 = length (m)

Inductance increases with more turns, larger area, higher permeability, and shorter length.

Filters

The output filter capacitor serves essentially the same function in the switching regulator as it does in the series-pass regulator; that is, it stores energy to be used by the load. Because the output ripple frequency is much higher, the output filter capacitor usually is a much smaller value than for the series-pass regulator.

The input filter must reduce the 60Hz or 120Hz ripple to acceptable values and/or it must keep the switching-frequency ripple from the input to keep the system stable and noise free. Switching power supplies can be a significant source of electromagnetic interference (EMI). Shielding and filtering are used to keep any generated EMI from escaping.

Oscillator and PWM Circuit

Refer again to *Figure 5-2* where the operation and waveforms of the oscillator and PWM were discussed. A convenient IC for both functions is the 556 dual timer. *Figure 5-6* shows the 556 interconnected as an oscillator and a PWM.

The first timer is used as a free-running oscillator, or multivibrator. The oscillation frequency is calculated by the following equation:

$$f = \frac{1.44}{(R_A + 2R_B)C_T}$$

The ON output pulse (C) from the PWM (the second timer) to the control element is triggered on by the constant frequency pulses (A) from the oscillator. When triggered, the PWM output goes high, turning on the control element. The output remains high for a time determined by the control voltage

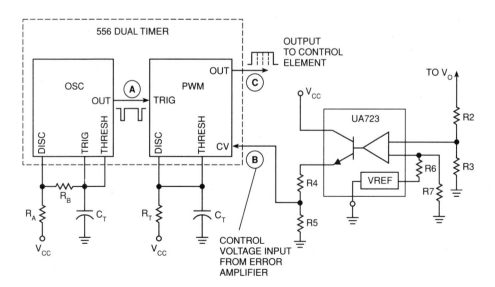

Figure 5-6. Oscillator, PWM and Error Amplifier

(B) from the error amplifier. When the ON time has expired, the PWM resets the output, forcing it low and turning off the control element. The PWM output remains low until it is triggered again by the oscillator. The time range of the ON pulse is controlled by R_T and C_T, the oscillator frequency by R_A and R_B and its C_T, and the trigger pulse width by R_B.

Error Amplifier and Voltage Reference

A convenient IC to use for the error amplifier and voltage reference is the 723 regulator described in *Figure 3-8*, and shown again in *Figure 5-6*. The sampling element, made up of R2 and R3, delivers a feedback voltage to one input of the error amplifier. The error amplifier compares the feedback voltage to the reference voltage, and produces a corresponding drive current for the 723's output transistor. The output resistors R4 and R5 develop the final control voltage connected to the PWM input. When using the 556 as a PWM, $R_4=0$ because the voltage range of $3.3-10.3V$ is a convenient range for the 556. If V_0 of the regulator is greater than the 7.15V reference voltage, the reference voltage can be connected directly to the inverting input of the error amplifier; however, this restricts the voltage swing of the control voltage. It is best to divide down the reference voltage with R_6 and R_7 to a lower voltage, and set the voltage on the inverting input from the sampling element to the same value.

DESIGN CHOICES

In any switching regulator design, certain design specifications are set for the design. Using these, the component values and design parameters are calculated and selected for the type of regulator used for the design.

Design Specifications

Output Voltage
What nominal output voltage(s) is required by the load? This sets V_O.

Load Current
How much current is required by the load? Sometimes this specification is for small load variations that may occur around a steady-state load current. If the current will not vary from a minimum load to full-load, some of the components may be smaller, or the efficiency higher, etc. The regulator must still be capable of supplying the full-load current.

Regulation
What percent regulation is required by the circuits or equipment to be powered? Some circuits need only 10%, while others need 1%.

Ripple Voltage
How much noise or ripple variation can the circuits or equipment tolerate? This specification determines the quality of filtering required on the regulator output.

Design Parameters

With knowledge of the specifications, the available input source, the available components, and the design flexibility, the regulator type can be selected and the design parameters determined. The following parameters usually are required:

  a. Duty cycle and ON time   d. Switching frequency
  b. Peak current   e. Output filter capacitance
  c. Coil inductance   f. Switching transistor and diode ratings.

Duty Cycle and ON Time

If a step-down regulator is used, in order to support an output of V_O from a higher input voltage, V_{IN}, the area under the two voltage-time curves must be equal as shown in *Figure 5-4a*. $V_{IN} \times t_{ON}$ must equal $V_O \times T$. Thus, the duty cycle, DC, and t_{ON} equations are:

$$\text{Duty Cycle} = \frac{t_{ON}}{T} = ft_{ON} = \frac{V_O}{V_{IN}} \qquad t_{ON} = \frac{V_O T}{V_{IN}} = \frac{V_O}{fV_{IN}}$$

The duty cycle and output voltage for three type regulators are as follows. The equation for V_O of the step-down regulator shows that V_O can be controlled by controlling the duty cycle.[1]

	Step-Down	Step-Up	Inverter
Duty Cycle	$\dfrac{V_O}{V_{IN}}$	$\dfrac{V_O - V_{IN}}{V_O}$	$\dfrac{\{V_O\}}{\{V_O\} + V_{IN}}$
Output V	$V_O = V_{IN}DC$	$V_O = \dfrac{V_{IN}}{1 - DC}$	$\{V_O\} = V_{IN}\dfrac{DC}{1 - DC}$
			$\{V_O\}$ is the value of $-V_O$ without the sign

Peak Current

The peak current in the switching regulator occurs when the inductor current is just enough to discharge the inductor completely and replace all the charge on the output filter capacitance at maximum load current. It is determined by the maximum load current and the fact that the inductor must be allowed to completely discharge ($I_L = 0$). The peak current varies with the different regulators as follows:

Step-Down	Step-Up	Inverter
$I_{pk} = 2\,I_{LOAD}$	$I_{pk} = 2\,I_{LOAD}\dfrac{V_O}{V_{IN}}$	$I_{pk} = 2\,I_{LOAD}\left[1 + \dfrac{\{V_O\}}{V_{IN}}\right]$

[1] Linear Technology Corporation, Application Note 19, C. Nelson, *LT1070 Design Manual.*

Coil Inductance

The voltage across the inductance, L, during the peak current change in time t_{ON} determines L. The general equations for the voltage and inductance is:

$$V = \frac{L\ I_{pk}}{t_{ON}} \qquad L = \frac{V\ t_{ON}}{I_{pk}}$$

For the three types of regulators, L is (disregarding transistor and diode drops):[2]

Step-Down	Step-Up	Inverter
$L = \dfrac{V_{IN} - V_O}{I_{pk}} t_{ON}$	$L = \dfrac{V_{IN}\ t_{ON}}{I_{pk}}$	$L = \dfrac{V_{IN}\ t_{ON}}{I_{pk}}$

Knowing the peak current, t_{ON}, and the voltage(s), the inductance can be calculated. The inductance value for continuous current in the inductor is chosen based on the *minimum* current that the switching regulator must handle—the lower the current, the larger the inductance. Calculate the I_{pk} current using the minimum current, and use this value to calculate the inductance.

Switching Frequency

The period, T, must be long enough to allow $t_{ON} + t_d$. Using the voltage equation above, the value of t_d that allows L to discharge completely must be:

Step-Down	Step-Up	Inverter
$t_d = \dfrac{L\ I_{pk}}{V_O}$	$t_d = \dfrac{L\ I_{pk}}{V_O - V_{IN}}$	$t_d = \dfrac{L\ I_{pk}}{\{V_O\}}$
$t_d = \dfrac{[V_{IN} - V_O]t_{ON}}{V_O}$	$t_d = \dfrac{V_{IN}\ t_{ON}}{V_O - V_{IN}}$	$t_d = \dfrac{V_{IN}\ t_{ON}}{\{V_O\}}$

Since $T = t_{ON} + t_d$, $1/T$ determines the maximum frequency.

For the project designs of Chapter 6, t_{ON} is between 10 and 20 microseconds, and the duty cycle is set at the midrange of the load current to be regulated, and thus, the frequency is set. t_{ON} will vary as the regulator controls the pulse width, getting narrower as the load decreases. Since the frequency remains constant, t_{OFF} increases and exceeds t_d, and the inductor, being completely discharged, goes into discontinuous operation. Therefore, if the PWM and error amplifier pulse width versus control voltage curve will allow it, t_{ON} and the frequency are best set when handling the lowest current.

Output Filter Capacitance

Increasing the charge on a capacitor increases the voltage across it; therefore, the I_C (and charge) curves of *Figures 5-4* would cause a ripple voltage as shown in *Figure 5-4c*. Since Q=CV, then C=Q/V.

[2] Motorola Semiconductor Technical Data MC34163/D "Power Switching Regulators," Motorola Inc., 1990.

As shown in *Figure 5-4*, the charge added or subtracted from C_F is the area under the I_C current-time curves. The largest charge deviation is used to calculate the value of C_F to limit the voltage change to the specified ripple voltage. With V_r representing the ripple voltage, C_F is:

$$C_F = \frac{\Delta Q}{V_r}$$

Using geometry to calculate the current-time area $(I \times t)$ and considering that the $+Q$ is the same for t_{ON} and t_d or t_{OFF}, the equations for C_F of the different types of regulators are as follows:[3]

Step-Down	Step-Up	Inverter
$C_F = \dfrac{(I_{pk} - I_L)^2}{2V_r I_{pk}} \, t_{ON} \dfrac{V_{IN}}{V_O}$	$C_F = \dfrac{(I_{pk} - I_L)^2}{2V_r I_{pk}} \, t_{ON} \dfrac{V_O}{V_O - V_{IN}}$	$C_F = \dfrac{(I_{pk} - I_L)^2}{2V_r I_{pk}} \, t_{ON} \dfrac{\{V_O\} + V_{IN}}{\{V_O\}}$

By substituting the voltage specification for V_r, the value of C_F can be calculated.

Ratings for D1 and Q1

As stated previously, the diode is usually a Schottky diode because of its excellent switching times and low forward voltage. The diode's PIV need not be greater than the larger of the input or output voltage. It must handle the peak current of the regulator.

The transistor must have low forward voltage and must have a breakdown voltage greater than any output-to-input difference, including polarity. It also must handle the peak current of the regulator, and have switching times much less than any t_{ON} times.

SUMMARY

The ability to build a switching power supply opens the door to a wide range of battery-powered projects that would not be practical with linear regulators. This chapter presented the basic operating principles that govern switching power supplies, the basic types—step-down, step-up, and inverter—and their circuits, the calculation of design parameters and the selection of components. Chapter 6 shows how to build some useful switching power supply projects.

---

[3] Texas Instruments Application Report SLVA001, J. Spencer, E.J. Tobaben, "Designing Switching Voltage Regulators with TL497A."

Switching Power Supply Projects

This chapter contains two switching power supply projects to demonstrate the basic concepts and fundamentals for designing a switching power supply, not to design the most elaborate and exotic supply. All the parts used are commonly available from Radio Shack.

The integrated circuits used clearly show the selection of major functions, such as oscillator frequency, PWM pulse widths for the trigger and turn-on time pulses, and error-amplifier reference voltage. Major semiconductor manufacturers are now concentrating on integrating all the control functions of the switching power supply for lower-wattage power supplies into one IC.

DESIGN PARTS, LIMITATIONS AND CAUTIONS

Table 6-1 lists specifications for the ICs, semiconductors and inductors used for the projects. These specifications place boundary conditions on the project designs.

Table 6-1. Project Parts and Their Design Limitations

Part	Function	Case	Design Limitations
723	Error Amplifier Reference Voltage Output Control Voltage	NO14	V_{CC} greater than 9.5V 15mA max current Voltage swing—3.3V-10.3V
556 Dual Timer	Oscillator for Sw f Pulse-Width Modulator (PWM)	NO14	Some instability below pulse width of 10µs
IRF511	Switching Transistor N-Ch MOSFET	TO-220	Max I_{DS}=3A, V_{DS}=60V V_{GS}=20V(+ or −), P_D=20W @ 25°C, derate 0.16W/°C
Diode	"Catch" diode-Schottky	A1	PIV=40V Max I_F=1A (Three diodes are used in parallel to provide I_F=3A to match IRF511)
MJE34 (TIP42)	PNP Power Transistor	TO-220	Max I_C=10A, V_{CE}=40V h_{FE}=20-100 @ 3A P_D=90W w/heat sink
2N3053	NPN Switching Transistor	TO-39	Max I_C=0.7A, V_{CE}=40V h_{FE}=50-250 @ 0.15A, P_D=1W
L	100µH RF choke		Rated at 2A, Core volume limited
L	Snap-together choke with 23 turns #22 magnet wire—200µH		Air gap can cause inductance variations

ICs

For the projects, $V_{IN} = +12V$ to meet the V_{CC} requirement of the 723. Feedback resistors are used to limit the gain of the 723, even so, some manufacturer's parts seem to cause the closed loop operating point to be right at the edge of the range of the voltage swing. Some 556 dual timers multiple trigger and cause misoperation. A calibration procedure has been setup in Chapter 7 to assure proper closed-loop operation. This procedure should be followed before applying power to the switching power supplies.

Transistor and Diode

Note that the power FET, IRF511, used for switching has a maximum $I_{DS}=3A$, while the maximum collector current of the MJE34 is $I_C=10A$. However, the design limitation is the peak inductor current that can be switched through the Schottky diodes. Even though it is not the best design technique, three Schottky diodes are operated in parallel to provide the same 3A maximum switching current as the IRF511. This is not normal design practice because, at the maximum peak current of 3 amps, the diode with the lowest V_F tends to "hog" more current than the other two, which exceeds its individual current rating. If a diode fails, it usually shorts because of excess current, which protects the other diodes in parallel. A 1N5822 Schottky diode is an alternate choice if the Radio Shack part is not available.

Inductor

Readily available inductors from Radio Shack, though not the most efficient inductors, provide practical design examples. The 100µH RF choke has a continuous current rating of 2A; however, switching 3A at the step-down duty cycle provides satisfactory operation. The snap-together choke core was glued together with steel-filled, conductive epoxy cement before the coil was wound on it. Nova Magnetics wound a special 170µH inductor for a step-down design. Its core and coil specifications are given in the Appendix, as well as a list of other magnetic materials manufacturers and distributors.

Maximum Load Current

The maximum load current that can be drawn from the project designed power supplies can be limited by:
1. The peak current in the inductor, which is limited by the maximum current that can be switched by the transistor and diodes.
2. The maximum current limit of the power supply providing the input voltage (V_{IN}).

Step-up designs demand high currents from the input supply due to the output high-voltage power transfer. The input current to a power supply is calculated using the following power-transfer equation:

$$I_{IN} = \frac{V_O \times I_{LOAD}}{V_{IN} \times Eff}$$

Caution

Switching power supplies require a minimum load. If operated with no load, the regulator controller will decrease the PWM turn-on time to very narrow pulses, causing very high voltages from the inductor which will destroy the switching transistor, diodes, and possibly the other ICs. In addition, the discontinuous inductor operation on turnoff causes unstable operation and subharmonic frequency generation. A minimum load current limit for the inductance used is specified for each project design. A commercially available switching power supply may have an internal active load that draws the minimum current from the supply if there is no load on the power supply terminals. When an external load is placed on the supply that exceeds the minimum current, the active load cuts out, and is effectively removed from the circuit.

PROJECT #1 — STEP-DOWN CONVERTER +12V TO +5V

Project #1 is a regulated +5V supply to power digital logic circuits in an environment, such as an automobile, where the only available power source is approximately 12V dc. The circuit schematic is shown in *Figure 6-1.*

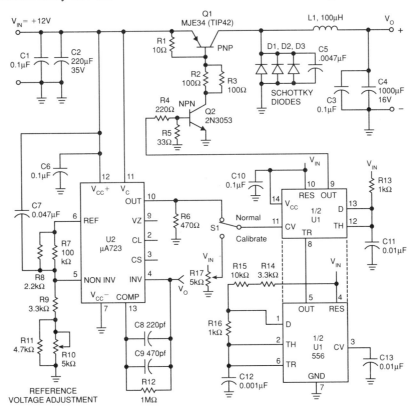

Figure 6-1. Step-Down Converter Circuit +12V to +5V

Design Specifications

V_{IN} = +12V

V_O = +5V

$I_{LOAD(MAX)}$ = 1.5A

$I_{LOAD(MIN)}$ = 300mA

Load Regulation = 2% for I_{LOAD} current
    range

Input Regulation = 2% for a ±5%
    input variation

Switching Frequency = approx. 50kHz

Ripple = 50mV at I_{LOAD} = 1A

Efficiency = 65% @ I_{LOAD} = 1A

Operating Ambient Temperature = 25°C

Input Voltage and Its Supply

The adjustable linear supply designed in Chapter 4, with a maximum load current limit of 1.5A, is used to supply V_{IN}. A regulated input supply simulates a +12V battery, eliminates V_{IN} variations, and makes it easier to build a successful, stable switching power supply. For an application other than +12V, the reader need only design an unregulated input supply to handle the load current variations expected.

Output Voltage

The +5V output is regulated to 2% over the load current range. The 2% regulation is maintained for an input voltage variation of ±5%.

Load Current

Using the input current equation and the equation for step-down peak current from Chapter 5 (calculations are shown in *Table 6-2*), the load current limit for the step-down design is 1.5A because the inductor peak current of 3A cannot be exceeded due to the Schottky diodes. The power supply efficiency was assumed to be 65%.

Table 6-2. Load Current Limits

	Limit		Limit
	I_{pk}	$I_{IN(max)}$	I_{LOAD}
$I_{LOAD} = \dfrac{I_{pk}}{2}$	3A	–	1.5A
$I_{LOAD} = \dfrac{V_{IN}I_{IN}Eff}{V_O}$	–	1.5A	2.34A

With transistors and diodes that have higher current ratings, the load current could increase to 2.34A before the I_{IN}=1.5A limit is reached.

Minimum load current is determined by the inductance used in the circuit. Since the design uses a 100µH choke and f =50kHz, the peak current at the minimum output load current can be determined from the equations in Chapter 5 as follows:

$$I_{pk} = \frac{V_{IN} - V_O}{L} \times t_{ON} = \frac{V_{IN} - V_O}{L} \times \frac{V_O}{fV_{IN}} = \frac{35}{60} = 0.58A = 580mA$$

Since $I_{pk} = 2I_{LOAD}$, minimum I_{LOAD} = 290mA.

The switching power supply regulates below the minimum current, but the inductor begins to operate in discontinuous mode, which generates more noise and is more difficult to stabilize.

Oscillator

The oscillator is one section of the 556 dual timer IC operating in an astable mode. C12 is the timing capacitor. The total series resistance of R14+R15 and R16 charge the capacitor during the time, t_1 shown in *Figure 5-2*, when the output is high, and R16 discharges the capacitor during the time, t_2, when the output is low.

The basic operating frequency of this supply is 50kHz, with a period, T, of approximately 20µs. The trigger pulse width, t_2, is slightly larger than 1µs. The trigger pulse width is initially calculated using the equation $t_2 = 0.693(R16)(C12)$, and the frequency using the equation $f = 1.44/[R14+R15+2(R16)](C12)$, as given in Chapter 5. According to the equations, $f = 94kHz$ using the final component values, and $t_2 = 0.693µs$. Internal delays of the 556 IC itself cause the equation to yield only approximate values. The final circuit values shown in *Figure 6-1* were determined experimentally using an oscilloscope to monitor the waveform.

Pulse-Width Modulator

The PWM half of the 556 is used in a "one-shot" configuration, where it produces a single positive pulse for every negative-going trigger pulse it receives from the oscillator. The width of the PWM output pulse depends on the values of C11 and R13, and on the control voltage. Without a control voltage on that input, $t_{ON} = 1.1(R13)(C11)$, or 11µs. The resistor and capacitor establish the range, and V_O from pin 10 of the 723 to the control voltage input to the 556 (pin 11) determines the exact pulse width. The output from the PWM section of the 556 (pin 9) drives the base of an NPN transistor in the switch circuit. Again, the equations predict approximate values; the values shown are determined experimentally.

Error Amplifier and Voltage-Reference

The switching power supply projects use the 723 regulator IC as the error amplifier and voltage reference. The 723's built-in current-limit circuit is not used. Pins 2 and 3 are left unconnected.

In this project, the output voltage is fed back directly to the inverting input of the error amplifier, pin 4, as illustrated in *Figure 6-1*. The reference voltage output is connected through a voltage divider (R7 ∥ R8) and (R9+R10+R11) to the non-inverting input of the error amplifier. R10 and R11 form a reference voltage adjustment used in the closed loop calibration procedure to adjust the output voltage of the regulator to +5V. The voltage of the non-inverting input to the 723 should be within 0.01V to 0.03V of +5V. A 0.05µF capacitor is connected between V_{IN} and the non-inverting input of the error amplifier to stabilize the error amplifier against noise feedback from V_{IN}.

To improve the regulation stability, a network consisting of a total of 690 pF (a 220pF and a 470pF) in parallel with a 1MΩ resistor is connected between the inverting error-amplifier input, pin 4, and the frequency-compensation input, pin 13. Both resistor and capacitor values were determined experimentally. Circuit layout can affect stability. False triggering and erratic output voltage under load are signs of unstable circuits. Keep all connecting leads as short as possible.

The error amplifier output from the 723 is an emitter follower. The emitter, pin 10, drives a 470Ω resistor connected to ground to supply the control voltage to pin 11 of the 556 PWM. This voltage ranges from 3.3V to 10.3V depending on the setting at the non-inverting reference input, the sample-voltage input, and the reference voltage. The midpoint output current condition of the design should set this voltage at about 7V. The collector of the internal output transistor, V_C, is tied to V_{CC} and the input voltage V_{IN}. A 0.1μF ceramic capacitor across V_{CC} reduces the effects of power-supply noise.

Inductor Choice

Using, f=50kHz, t_{ON} is determined from the duty-cycle equation in Chapter 5 as:

$$t_{ON} = \frac{V_O}{fV_{IN}} = \frac{5}{50 \times 10^3 \times 12} = 8.33\mu s$$

The inductance required for a peak current of 3A is calculated from the step-down equation in Chapter 5, as follows (f=50kHz):

$$L = \frac{12 - 5}{3} \times \frac{5}{50 \times 10^3 \times 12} = 19.4\mu H$$

An inductance of 19.4μH is the minimum inductance that can be used at a peak current of 3A at 50kHz. The minimum current of 290mA for continuous-mode operation using the 100μH choke was calculated using the same equation. 100μH is used so the minimum current can be at least 300mA. If a lower minimum current is required, a larger inductance must be used. The size of the inductor may affect the overall size of the power supply in many designs.

Switching Transistor Choice

An MJE34 PNP power transistor is chosen as the switch because of its low $V_{CE(SAT)}$. The PNP transistor has its emitter tied to V_{IN}=12V. It is turned on by pulling 150mA of base current through a 50Ω resistor to ground. To ensure a rapid turn-off time, a 10Ω resistor is connected between the base and emitter, providing approximately 70mA of turn-off current. The total base plus turn-off current is limited to approximately 220mA by the series 50Ω resistor, composed of two 100Ω, 1W resistors in parallel for extra power handling capability.

Since the output waveform of the PWM is high during the on time of the switch, the signal needs to be inverted from the PWM before driving the base of the MJE34. A small-signal NPN transistor, the 2N3053, inverts the PWM output and drives the base of the MJE34. The PWM output drives approximately 30 mA into the 2N3053 base through a series 220Ω current-limiting resistor. A 33Ω resistor from the base to ground provides 21mA of turn-off current to ensure rapid turn-off time.

"Catch" Diode Choice

The Schottky-barrier diode is rated at 40 volts peak repetitive reverse voltage, and will handle an average forward current of 1A. Three diodes are used in parallel to provide 3A. This is a weak link in the design, but was done to demonstrate that the rest of the circuitry could handle 1.5A.

Output and Input Filters

To keep the output ripple voltage (V_r) below 50mV, the minimum output-capacitor value is calculated using the step-down equation for C_F from Chapter 5. With $I_{pk}=3A$, $I_L=1.5A$ and $t_{ON}=8.33\mu s$, the minimum C_F for this supply is:

$$CF = \frac{(3-1.5)^2}{2 \times 0.05 \times 3} \times 8.33 \times 10^{-6} \times \frac{12}{5} = 150\mu F$$

A small size, 1000μF, 16V electrolytic capacitor with radial leads, almost 10 times the size needed, is selected for C_F. A 0.1μF, 50V ceramic capacitor is placed in parallel to bypass high-frequency noise.

A 220μF, 35V input capacitor acts as an input filter. It is necessary in a switching power supply because the switching action of the transistor can induce a hefty ripple voltage on the input supply voltage.

Construction Details

The project is built on a 1½" × 6" strip of perfboard (RS# 276-1396). This allows the same circuit layout of *Figure 6-3* to be included on one end of a circuit board for an application that is to be powered by the supply.

Component Layout

The component layout, while originally hand wired on perfboard, can be used as a guide for making a printed circuit board using a PC-board kit (RS# 276-1530). As shown in the photographs in *Figure 6-2,* power and ground conductors were made from 20-gauge hookup wire, with wire-wrap posts used to connect between the circuit side and component side of the board. Smaller wired connections were made either with wire-wrap wire or by bending the component leads into the desired shape after insertion into the board.

Figure 6-3 shows a parts-placement and wiring diagram from the component side and the circuit side. The component leads and wiring on each side of the board are shown as solid lines. The parts used are listed in *Table 6-3.*

Figure 6-2. Step-Down Switching Power Supply +12V to +5V

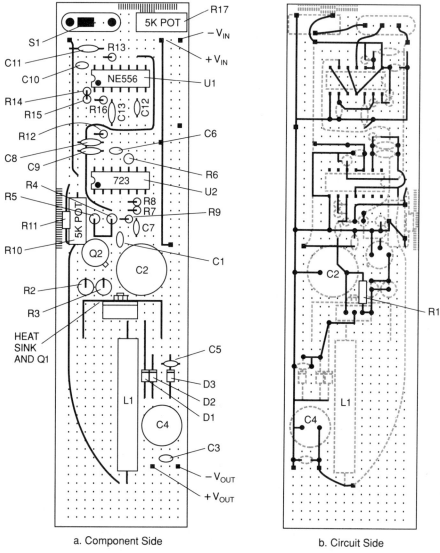

a. Component Side

b. Circuit Side

Figure 6-3. Parts-Placement and Wiring Diagram

Operation and Testing

A calibration procedure is included in Chapter 7. Check out your total circuit carefully and assure that all connections are correct. *Make sure that you apply no power to your circuit until you perform the calibration procedure.* Because of component tolerances, values stated in a design may vary due to the combination of tolerances present in the circuit.

Table 6-3. Parts List for Step-Down Converter

Description	Quantity	Reference Designator
Perfboard, 4″ × 6″	1	
Hookup wire, 20 gauge	1	
Wire-wrap wire	1	
Wire-wrap posts	1	
SPDT Slide Switch	1	S1
IC, dual timer, NE556	1	U1
IC, voltage regulator, 723	1	U2
Transistor, PNP power, MJE34 Q1	1	Q1
Transistor, NPN, 2N3053	1	Q2
Diode, Schottky 1A, 40V	3	D1, D2, D3
RF Choke	1	L1
Resistor, 10Ω, ¼W	1	R1
Resistor, 100Ω, 1W	2	R2, R3
Resistor, 220Ω, ¼W	1	R4
Resistor, 33Ω, ½W	1	R5
Resistor, 470Ω, ½W	1	R6
Resistor, 100kΩ, ¼W	1	R7
Resistor, 2.2kΩ, ¼W	1	R8
Resistor, 3.3kΩ, ¼W	2	R9, R14
Potentiometer, 5kΩ	2	R10, R17
Resistor, 4.7kΩ, ¼W	1	R11
Resistor, 1MΩ, ¼W	1	R12
Resistor, 1kΩ, ¼W	2	R13, R16
Resistor, 10kΩ, ¼W	1	R15
Capacitor, .1µF, 50V ceramic	4	C1, C3, C6, C10
Capacitor, 220µF, 35V electrolytic	1	C2
Capacitor, 1000µF, 16V electrolytic	1	C4
Capacitor, .0047µF, 50V ceramic	1	C5
Capacitor, .047µF, 50V ceramic	1	C7
Capacitor, 220pF, 50V ceramic	1	C8
Capacitor, 470pF, 50V ceramic	1	C9
Capacitor, .01µF, 50V ceramic	2	C11, C13
Capacitor, .001µF, 50V ceramic	1	C12

NOTE: All 1/4W resistors are ±5% tolerance
      All 1/2W resistors are ±10% tolerance

PROJECT #2 — STEP-UP CONVERTER +12V to +24V

An application that is well served by a switching regulator is when a dc output voltage is required that is higher than an input dc voltage. Project #2 is a regulated +24V supply that has an input voltage of +12V. It is a design that demonstrates a basic step-up converter circuit, and shows some of its advantages and limitations. The schematic is shown in *Figure 6-4.*

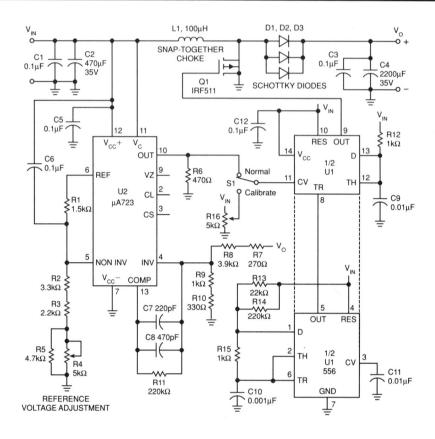

Figure 6-4. Step-Up Converter Circuit +12V to +24V

Design Specifications

V_{IN} = +12V
V_O = +24V
$I_{LOAD(MAX)}$ = 750mA*†
$I_{LOAD(MIN)}$ = 200mA
Load Regulation = 2% for I_{LOAD} current range

Input Regulation = 2% for a ±5% input variation
Switching Frequency = 41kHz
Ripple = 100mV at I_{LOAD} = 0.6A
Efficiency = 90%
Operating Ambient Temperature = 25°C

*limited to 675 mA by input supply (90% eff.)
†limited by 3A maximum switching current (100% eff.)

Input Voltage and Its Supply

The input power supply is again the linear adjustable supply of Chapter 4 with its 1.5A maximum load current limit.

Output Voltage

The +24V output voltage is regulated to within 2% over the output current range, and over a ±5% variation in input voltage.

Load Current

Since the input supply can provide no more than 1.5A, it limits the load current that can be drawn from the step-up converter. As a result, the maximum load current is limited to 0.675A, assuming 90% efficiency for the power supply. The calculations are shown in *Table 6-4*. If an input power supply is used that can supply more than 1.5A, then the load current could increase to 0.75A before the peak inductor current limit of 3A is reached.

Table 6-4. Load Current Limit

		Limit		Limit
		I_{pk}	$I_{IN(max)}$	I_{LOAD}
$I_{LOAD} = \dfrac{V_{IN}I_{pk}}{2V_O}$		3A	–	0.75A
$I_{LOAD} = \dfrac{V_{IN}I_{IN}Eff}{V_O}$		–	1.5A	0.675A

The inductance used for project #2 is a 200µH coil wound with 18 turns of #22 magnet wire on the snap-together toroidal core. The minimum load current for this inductance, based on a frequency of approximately 41kHz, and a duty cycle of 0.5, is 180mA based on the following calculations:

$$I_{pk} = \frac{V_{IN}}{L} \times t_{ON} = \frac{12}{200 \times 10^{-6}} \times 12 \times 10^{-6} = 0.72A$$

Substituing 0.72A in the I_{LOAD} equation in *Table 6-4*, $I_{LOAD\ (max)}$ = 180mA.

Oscillator

The 556-based oscillator/PWM circuit is identical to that of the step-down converter with some of the values changed. The values for the step-up converter are as follows:

R14 in parallel with R13 = 20kΩ   R15 = 1.0kΩ   C10 = 0.001µF

The resultant frequency is approximately 41kHz. The trigger pulse width is slightly larger than 1µs.

Pulse-Width Modulator

The output from the PWM section of the 556 (pin 9) drives the gate of the IRF511 power FET. The values for the resistor and capacitor are as follows:

C9 = 0.01µF   R12 = 1.0kΩ

The t_{ON} time for median pulse width is 12µs.

Error Amplifier and Voltage-Reference Choice

The error amplifier and voltage reference is provided by the 723 regulator as in the step-down converter. Since the output voltage that must be sampled is greater than 7 volts, the error amplifier circuit shown in *Figure 6-4* differs slightly from that of the step-down converter. The sampling element is an additional voltage divider from V_O with the junction of R8 and R9 providing the sampled voltage to the inverting input of the 723. When the output is at 24 volts, the voltage at the inverting input is from 5.85 to 5.95 volts. The reference voltage is divided down to approximately 5.90 volts at the non-inverting input by the voltage divider formed by resistors R1 through R5. Three circuit design techniques assure regulator stability. First, a 0.1µF capacitor is connected from V_{IN} to the non-inverting input, pin 5. Second, a 690pF capacitor in parallel with a 220kΩ resistor are connected between the error-amp input, pin 4, and the frequency-compensation input, pin 13. Third, a 0.1µF ceramic capacitor is placed from V_{cc} to ground to reduce the effects of power-supply noise.

Pin 11, the collector of the internal output transistor is tied to V_{IN}, and the emitter drives a 470Ω resistor connected to ground. This is the control voltage that varies the PWM output.

Inductor Choice

I_{pk} is again limited by the Schottky diodes to 3A, and is used to determine the minimum inductance as follows:

$$L = \frac{V_{IN}}{I_{pk}} \times t_{ON} = \frac{12}{3} \times 12 \times 10^{-6} = 48\mu H$$

Only 48µH are required when the power supply operates at maximum inductor current, but the 200µH value is chosen to allow the regulator to operate at a minimum current of 180mA. If a lower minimum current is required, the inductor would have to be larger.

Coil Fabrication

The two ferrite halves of the snap-together core are glued together using a steel-filled, conductive epoxy cement. Electrical tape is wrapped around the core before the #22 magnet wire is wound evenly on it. The completed inductor is shown in *Figure 6-6c.*

Switching Transistor Choice

An IRF511 power field-effect transistor (FET), rated at $I_{DS(max)}$=3A continuous, is used as the switching transistor in the step-up design. The breakdown voltage is 60V, well above the levels encountered in this supply. The power FET has a maximum forward voltage drop of 1.8V at the peak current of 3A. The input capacitance of a power FET is relatively large. The 556 has a totem-pole output stage that rapidly charges and discharges the IRF511 input capacitance, ensuring fast turn-on and turn-off times.

"Catch" Diode Choice

Three Schottky barrier diodes are used in parallel to handle a peak inductor current of 3A, even though the maximum load current of 1.5A from V_{IN} limits the maximum inductor current to 2.7A.

Output and Input Filters

To meet the goal of an output ripple voltage (V_r) below 100mV, the minimum output-capacitor value (C_F) is calculated using the step-up equation in Chapter 5. $t_{ON}=12\mu s$, $I_{pk}=2.7A$, and $I_L=0.675A$.

$$C_F = \frac{(2.7 - 0.675)^2}{2 \times 0.1 \times 2.7} \times 12 \times 10^{-6} \times \frac{24}{24 - 12} = 182\mu F$$

The actual output capacitor is 10 times this value to minimize output ripple. The 2200µF, 35V, electrolytic capacitor with axial leads has a size, even for such a large capacitance, that fits nicely on the circuit board. Again a 0.1µF, 50V, ceramic capacitor in parallel bypasses high-frequency noise.

Construction Details

This project is again built on 1½"×6" strip of perfboard, and is nearly identical in layout and construction to the step-down converter. The same interconnection materials were used.

Component Layout

The component layout is shown in the photographs in *Figure 6-5*.

a. Component Side

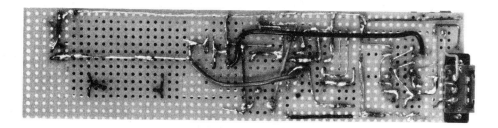

b. Circuit Side

Figure 6-5. Step-Up Switching Power Supply +12V to +24V

Figure 6-6 shows a parts-placement and wiring diagram from the component and the circuit side. The component leads and wiring on each side of the board are shown as solid lines. The parts used are listed in *Table 6-5*.

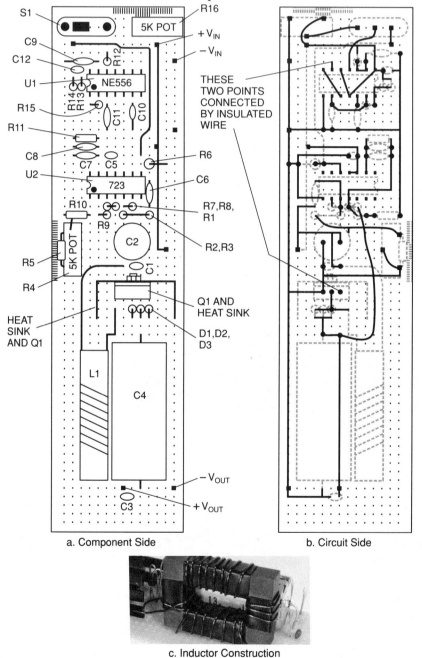

a. Component Side

b. Circuit Side

c. Inductor Construction

Figure 6-6. Parts-Placement and Wiring Diagram

Operation and Testing

A calibration procedure is included in Chapter 7. Check out your total circuit carefully and assure that all connections are correct. *Make sure that you apply no power to your circuit until you perform the calibration procedure.* Because of component tolerances, values stated for a design may vary due to the combination of tolerances present in the circuit.

Table 6-5. Parts List for Step-Down Converter

Description	Quantity	Reference Designator
Perfboard, 4″ × 6″	1	
Hookup wire, 20 gauge	1	
Magnet wire, 22 gauge	1	
Wire-wrap wire	1	
Wire-wrap posts	1	
SPDT Slide Switch	1	S1
IC, dual timer, NE556	1	U1
IC, voltage regulator, 723	1	U2
FET, Power, IRF511	1	Q1
Diode, Schottky 1A, 40V	3	D1, D2, D3
Core, Snap Together	1	L1
Resistor, 1.5kΩ, ½W	1	R1
Resistor, 3.3kΩ, ¼W	1	R2
Resistor, 2.2kΩ, ¼W	1	R3
Potentiometer, 5kΩ, ¼W	2	R4, R16
Resistor, 4.7kΩ, ¼W	1	R5
Resistor, 470Ω, ½W	1	R6
Resistor, 270Ω, ¼W	1	R7
Resistor, 3.9kΩ, ¼W	1	R8
Resistor, 1kΩ, ¼W	3	R9, R12, R15
Resistor, 330Ω, ¼W	1	R10
Resistor, 220kΩ, ¼W	1	R11, 14
Resistor, 22kΩ, ¼W	1	R13
Capacitor, .1µF, 50V ceramic	4	C1, C3, C5, C6
Capacitor, 470µF, 35V electrolytic	1	C2
Capacitor, 2200µF, 35V electrolytic	1	C4
Capacitor, 220µF, 50V ceramic	1	C7
Capacitor, 470µF, 50V ceramic	1	C8
Capacitor, .01µF, 50V ceramic	2	C9, C11
Capacitor, .001µF, 50V ceramic	1	C10

NOTE: All 1/4W resistors are ±5% tolerance
　　　All 1/2W resistors are ±10% tolerance

Summary

This chapter concludes the power supply projects. Switching power supply stability can be very sensitive to layout and to the ICs used. Make sure that the calibration procedure in Chapter 7 is followed to assure the regulators will operate properly. In addition, Chapter 7 discusses measurement techniques, presents data on all the project power supplies, and provides troubleshooting tips.

Measurements, Calibration and Troubleshooting

In this final chapter, results of performance measurements on the power-supply projects from Chapters 4 and 6 are listed, along with troubleshooting tips and a calibration procedure for the switching power supplies.

TYPES OF MEASUREMENTS

The measurements made on the linear supplies were load regulation, power output, percent regulation, and rms ripple voltage. The measurements made on the switching power supplies included input regulation, conversion efficiency, ripple voltage, control voltage, t_{ON}, and the sample and reference voltages under various loads. The results are provided in a series of tables. The percent of load regulation is calculated using the equation given in Chapter 4.

LINEAR POWER SUPPLIES

Table 7-1. +5V Series-Pass Power-Supply Measurements

V_o V	I_o A	P_o W	$V_{r(rms)}$ mV	%Reg %
4.99	0	0	<1	–
4.98	.05	.25	<1	0.4%
4.87	.49	2.37	<1	2.6%
4.74	.95	4.49	<1	5.2%
4.61	1.40	6.44	<1	8.2% *

*Not to be operated continuously unless cooled with fan

Table 7-2. ±12V Series-Pass Power-Supply Measurements

V_o+ V	V_o- V	I_o A	P_o W	$V_{r(rms)}$ mV	%Reg %
11.81	11.82	0	0	<1	–
11.80	11.81	.05	0.59	<1	0.08%
11.79	11.80	.6	7.1	<1	0.16%
11.78	11.78	1.5	17.7	<1	0.25%

Table 7-3. Adjustable Power-Supply Measurements

V_o V	I_o A	P_o W	$V_{r(rms)}$ mV	%Reg %
3.0	0	0	<3	–
3.0	.05	.15	<3	–
3.1	.5	1.5	<3	0.33%
6.0 (6.11 @ I_o = 0)	1.5	9	<3	1.8%
14.99 (15.0 @ I_o = 0)	.05	.75	<5	0.06%
14.98	.5	7.5	<5	0.13%
14.97	1.5	22.5	<5	0.20%
25 (25.0 @ I_o = 0)	.05	1.25	<10	–
24.9	.5	12.5	<10	0.40%
24.8	1.5	37.5	<10	0.81%

SWITCHING POWER SUPPLIES

The switching power supplies in Chapter 6 used readily available parts, but some of the parts are capable of greater precision than others, therefore, adjustments are provided to overcome these variations. First, a potentiometer allows the reference voltage applied to the non-inverting input (pin 5) of the 723 IC to be varied. Second, a switch allows the control loop to be opened, and with a voltage from another potentiometer, to directly vary the PWM output pulse width. These adjustments, used with the calibration procedure that follows, allow the switching power supplies to be precisely adjusted to the correct output voltage, regardless of component variations.

Switching Power Supply Calibration Steps

1. Disconnect the power supply providing V_{IN}.
2. Place S1 in the Calibrate (open-loop) position, toward the open-loop potentiometer (R17 for the +5V supply, R16 for the +24V supply). This connects the wiper tap of the potentiometer to the PWM control-voltage input.
3. Set the potentiometer to the middle of its adjustment range.
4. Connect a load resistor across the output of the supply that will draw approximately 500mA (10Ω for +5V, 50Ω for +24V).
5. Connect a voltmeter across the power supply output to monitor V_O.
6. Turn on the power supply for V_{IN}.
7. Adjust the open-loop potentiometer to obtain the desired V_O— +5V for the step-down converter, +24V for the step-up converter.
   CAUTION: Turning the open-loop potentiometer should produce a smooth, even change in V_O. If the V_O change is erratic, replace the 556 and repeat the procedure. Some 556s double trigger, causing erratic operation.
8. With V_O set to the correct level, measure the voltage at pin 4 of the 723, and record it.
9. Measure the voltage at pin 5 of the 723, and adjust the reference-voltage adjustment potentiometer until the voltage at pin 5 equals the voltage at pin 4. If, on the step-up converter, the potentiometer does not have enough adjustment range to bring the pin-5 voltage down to the level at pin 4, replace R2 with a wire (shorts R2) and repeat this step.
10. Measure the control-voltage output of the 723 at pin 10. It should be within the 3V–10V linear range of the 723. If the voltages on pins 4 and 5 differ by more than 0.1V, the 723 output will be either 3V or 10V, depending on whether pin 4 or pin 5 is greater. For the supply to operate properly, the voltages at pins 4 and 5 must be the same. If they are, and the pin-10 voltage is outside the linear range, replace the 723 and repeat the procedure.
11. Turn off V_{IN}, and place S1 in the Normal (closed loop) position which connects pin 10 of the 723 to pin 11 of the 556.
12. The supply is now calibrated and should regulate V_O to the desired voltage when V_{IN} is turned on. Slightly trimming the reference-voltage adjustment potentiometer may result in greater V_O accuracy.

Table 7-4 +5V Step-Down Converter Measurements

V_{in} V	I_{in} mA	P_{in} W	V_o V	I_o A	P_o W	Eff %	$V_{r(rms)}$ mV	C_v V	t_{ON} μs	V_{samp} V	V_{ref} V
12	180	2.2	4.93	0.16	0.87	36.4	15	3.0	5	4.93	4.90
12	385	4.6	4.93	0.49	2.4	52	15	6.2	7.2	4.93	4.90
12	580	7	4.93	0.98	4.8	69	15	6.7	7.8	4.93	4.90
12	790	9.5	4.93	1.47	7.2	76	16	7.1	8.2	4.93	4.90

Table 7-5 +24V Step-Up Converter Measurements

V_{in} V	I_{in} A	P_{in} W	V_o V	I_o mA	P_o W	Eff %	$V_{r(rms)}$ mV	C_v V	t_{ON} μs	V_{samp} V	V_{ref} V
12	.52	6.2	24.0	203	4.9	78.5	44	7.9	11.0	5.73	5.7
12	1.01	13.2	24.0	480	11.5	87	71	8.2	11.9	5.73	5.7
12	1.30	15.6	23.8	595	14.2	91	150	8.2	12	5.73	5.7

TROUBLESHOOTING TIPS—INITIAL OPERATION

There are five major reasons that a power supply circuit doesn't operate initially:

1. Wrong Circuit Connection—Check the circuit carefully to make sure all connections are correct. The switching power supply circuits are somewhat detailed, and thus, more difficult to build.

2. Bad Connections—It is easy to have a cold solder joint so that a connection is open; reheat the suspected joints. Or solder may have run between connections to produce a short; especially on ICs. Use an ohmmeter or continuity meter to check for shorts.

3. Wrong Component Values—Assure that all component values are correct. A common cause is that the color coding on resistors was read incorrectly.

4. Bad Components—Integrated circuits, transistors and diodes are very sensitive to heat, static electricity, and conditions that exceed specifications. They may have been damaged when soldered if heated too long, or because a built-up static charge discharged through it. Or due to wrong circuit connections, voltage, current or power specifications may have been exceeded. Remember, the linear regulator ICs will shut down if they are asked to supply excessive current or get too hot. A short in the circuit could cause this condition and make it appear that the regulator is bad. Look at the problem and analyze it carefully.

5. Circuit Oscillations—When a circuit oscillates, a signal is being feedback from output to input to provide positive feedback to cause the resultant output to get larger rather than negative feedback to keep the output under control. The cause is that there is too much gain in the circuit or the feedback signal time relationship (phase) is wrong. Clean circuit layout and short leads are very important to keep circuits from oscillating. A 0.1μF capacitor to ground or to the power rail is a good tool to try to stabilize oscillations. Connect it to the various circuit points by trial and error. If the circuit stabilizes, see if a layout change or amplifier gain reduction might solve the problem.

The feedback loop may have to be opened so that the circuit can be adjusted to proper operating conditions before the loop can be closed. Use the switching power supply calibration procedure as an example of how this is done.

It is suggested that voltage measurements be made around the power supply circuit once it is operating to use for troubleshooting if the power supply were to fail.

TROUBLESHOOTING TIPS—OPERATIONAL FAILURE

After a power supply has been operating properly and then fails, here are some common causes:

1. Fuses—A fuse has blown. Replace the fuse.
2. Exceeding Specifications—Sometime during the use of the power supply, the specifications were exceeded. If a short is placed on the switching power supplies of Chapter 6, the switching current of the diodes will be exceeded, and the diodes will be damaged. In this case, the diodes usually short rather than open. If the supplies are allowed to operate without a load, high voltage will destroy the diodes and all the ICs. In this case, the diodes usually open. The only solution is to replace components.
3. A Connection Has Opened or Shorted—Make voltage and continuity or resistance measurements to isolate the problem. Repair the faulty connection. Here is where a chart of voltages around the circuit, taken when the circuit is operating properly, is very helpful.
4. A Component Has Failed—Resistors may change in value, capacitors and transformers usually short, switches and contacts become corroded. ICs and other semiconductors open because of excessive temperature cycling, short because of excessive voltage or current, or burst open because of excessive heat.

VOLTAGE, CURRENT AND RESISTANCE MEASUREMENTS

Remember, when making voltage measurements, measure across a component, or an input or output. When measuring current, turn off the power, break into the circuit, and put the multitester in series with the circuit component. When measuring a component's resistance, turn off the power and disconnect one end of the component from the circuit. Leaving the component in the circuit may damage the meter or may cause errors in readings. For anyone needing help with multitester measurements, Prompt Publications has a book titled *VOM and DVM Multitesters for the Hobbyist and Technician* that shows how to apply a multitester.

SUMMARY

In this book, basic concepts and fundamentals are presented, including design guidelines, component selection and why, and actual practical designs detailed so they can be built. The objective was to promote understanding and have success by applying it. We hope we have met our goal.

Appendix

Inductors for Switching Power Supplies of Chapter 6

Specifications:
Core: 0.687″ OD
  0.375″ ID
  0.187″ H (Thick)

Powered Iron Toroid
Permeability = 75
With 42 turns of #22 magnet wire
Switching Inductance @ 0.5A = 170μH

Supplier:
Nova Magnetics, Inc.
1101 E. Walnut St.
Garland, TX 75040
Part No. 9595-09-0057

Other Suppliers of Same Core:
Micrometals, Inc.
1190 N. Hawk Circle
Anaheim, CA 92087-1788
Part No. T68-26

Pyroferric International, Inc.
200 Madison Street
Toledo, IL 62468
Part No. PT680-75

Permacor, Inc.
9540 S. Tulley Ave.
Oak Lawn, IL 60453
Part No. P75-T17-9-5

Some Other Magnetic Materials Manufacturers or Distributors

South Haven Coil, Inc.
P.O. Box 409
Blue Star Highway
South Haven, MI 49090

Phillips Components
5083 Kings Highway
Saugerties, NY 12477

Pulse Engineering
7250 Convoy Court
San Diego, CA 92111

Coil craft
1102 Silver Lake Rd.
Cary, IL 60013

Renco Electronics, Inc.
60 Jeffryn Blvd. East
Deer Park, NY 11729

Howard W. Sams & Company does not endorse or in any way imply preference for, or warrant, any of the above listed suppliers.

Glossary

AC line: Alternating-current power distribution line. Typically 110V-125V at 60Hz in the United States.

Alternating current: An electrical current (produced by a voltage) that periodically changes in magnitude and direction.

Ampere: The unit of measurement for electrical current in coulombs (6.25×10^{18} electrons) per second. There is one ampere of current in a circuit that has one ohm resistance when one volt is applied to the circuit. See Ohm's law.

Amplifier: An electrical circuit designed to increase the current, voltage or power of an applied signal.

Capacitor: A device made up of two metallic plates separated by a dielectric (insulating material). Used to store electrical energy in the electrostatic field between the plates. It produces an impedance to an ac current, and opposes changes to the voltage across it.

Circuit: A complete path that allows electrical current from one terminal of a voltage source to the other terminal.

Closed loop: When used in reference to power supplies, the completed circuit of the control loop wherein a portion of the output is fedback to the input to accomplish the control of the output.

Control loop: The circuit consisting of control device, power-supply output, sampling circuit, feedback signal, error amplifier, and control voltage. See regulator action discussion in Chapter 3.

Conversion efficiency: The percentage of input power that a power supply converts to useful energy at its output.

Current (I): The flow of charge (electrons) measured in amperes. See ampere.

Direct Current(dc): Current in which the charge (electrons) flows in only one direction.

Electromagnetic interference (EMI): Disruption of the proper operation of a radio receiver or other electronic circuit caused by electromagnetic radiation (noise) from another circuit. This interfering noise may be transmitted through the air or conducted as power-supply noise.

Farad (F): The basic unit of capacitance. A capacitor has a value of one farad when it can store one coulomb of charge with one volt across it.

Feedback: An electrical signal from a later processing stage in a regulated power supply that gives an earlier stage the information it needs to properly do its task, e.g.,the sampling circuit provides a feedback voltage to the error amplifier.

Filter: A circuit element or group of components which passes signals of certain frequencies while blocking signals of other frequencies.

Frequency: The number of complete cycles of a periodic waveform during one second, expressed as hertz.

Ground: Refers to a point of (usually) zero voltage, and can pertain to a power circuit or a signal circuit.

Hertz: A unit of frequency equal to one cycle per second, named after German physicist H.R. Hertz.

Impedance(Z): The opposition (measured in ohms) of circuit elements to alternating current. The impedance includes both resistance and reactance.

Inductance(L): The capability of a coil to store energy in a magnetic field surrounding it. It produces an impedance to an ac current, and opposes changes in current through it.

LC filter: Filter composed of both capacitors (usually paralleled with the load) and inductors (in series with the load).

Line voltage: See AC line.

Load/line regulation: The maximum output-voltage variation allowed by a regulated power supply in response to load-current or input (line) variations.

Ohm (Ω): A unit of electrical resistance, reactance or impedance.

Ohm's law: A basic law of electric circuits. It states that the current I in amperes in a circuit is equal to the voltage E in volts divided by the resistance R in ohms; thus, $I = E/R$.

On time: The portion of the switching waveform during which the control-switching device conducts.

Off time: The portion of the switching waveform during which the control-switching device is not conducting.

Peak: The maximum amplitude of a voltage or current. For a sine-wave ac, $V_{peak} = 1.414\ V_{RMS}$.

Peak-to-Peak: The magnitude of the difference between the maximum positive and negative peaks of a voltage or current.

Period: For electrical circuits, the length of time required for one cycle of a periodic wave.

Phase: The angular or time displacement between the voltage and current in an ac circuit.

Polarity: In circuits, the description of whether a voltage is positive or negative with respect to some reference point.

Potentiometer: A variable resistance with a wiper mounted on a rotating shaft.

Power(P): The rate at which energy is used (voltage times current.)

Reactance: The opposition that a pure inductance or pure capacitance provides to a current in an ac circuit.

RMS: An acronym for root mean square. The RMS value of an alternating current produces the same heating effect in a circuit as the same value of direct current. For a sine-wave, $V_{RMS} = 0.707\ V_{peak}$.

Sine Wave: A waveform of an line alternating current or voltage. Its instantaneous magnitude is proportional to the sine of the angle of rotation of the coil generating the voltage.

Voltage or Volt(V): The unit of electromotive force that causes current when included in a closed circuit. One volt causes a current of one ampere through a resistance of one ohm. See Ohm's law.

Watt(W): A unit of electrical power. It is the use of one joule of energy per second. One volt times one ampere equals one watt. See Power.

Index

BUSINESS REPLY MAIL

FIRST CLASS MAIL   PERMIT NO. 1317   INDIANAPOLIS IN

POSTAGE WILL BE PAID BY ADDRESSEE

HOWARD W. SAMS & COMPANY

2647 WATERFRONT PKY EAST DR

INDIANAPOLIS IN 46209-1418

☞ **Dear Reader:** *We'd like your views on the books we publish.*

PROMPT® Publications, an imprint of Howard W. Sams & Company, is dedicated to bringing you timely and authoritative documentation and information you can use.

You can help us in our continuing effort to meet your information needs. Please take a few moments to answer the questions below. Your answers will help us serve you better in the future.

1. Where do you usually buy books?_____

2. Where did you buy this book?_____

3. Was the information useful? _____

4. What did you like most about the book? _____

5. What did you like least?_____

6. Is there any other information you'd like included?_____

7. In what subject areas would you like us to publish more books?

   (Please check the boxes next to your fields of interest.)

   ❑ Amateur Radio   ❑ Computer Software

   ❑ Antique Radio and TV   ❑ Electronics Concepts/Theory

   ❑ Audio Equipment Repair   ❑ Electronics Projects/Hobbies

   ❑ Camcorder Repair   ❑ Home Appliance Repair

   ❑ Computer Hardware   ❑ TV Repair

   ❑ Computer Programming   ❑ VCR Repair

8. Are there other subjects not covered in the checklist that you'd like to see books about?

   _____

9. Comments _____

   _____

   _____

Name_____

Address_____

City_____ State/Zip_____

Occupation _____ Daytime Phone_____

Thanks for helping us make our books better for all of our readers. Please drop this postage-paid card in the nearest mailbox.

For more information about PROMPT®Publications, see your
authorized Sams PHOTOFACT® distributor.
Or call 1-800-428-7267 and ask for Operator MP2.

PROMPT.
PUBLICATIONS

Imprint of Howard W. Sams & Company
2647 Waterfront Parkway East Drive, Indianapolis, IN
46214-2041